Zukunftstechnologien für den multifunktionalen Leichtbau

Series Editor
Open Hybrid LabFactory e.V.

Ziel der Buchreihe ist es, zentrale Zukunftsthemen und aktuelle Arbeiten aus dem Umfeld des Forschungscampus Open Hybrid LabFactory einer breiten Öffentlichkeit zugänglich zu machen. Es werden neue Denkansätze und Ergebnisse aus der Forschung zu Methoden und Technologien zur Auslegung und großserienfähigen Fertigung hybrider und multifunktionaler Strukturen vorgestellt. Insbesondere gehören neue Produktions- und Simulationsverfahren, aber auch Aspekte der Bauteilfunktionalisierung und Betrachtungen des integrierten Life-Cycle-Engineerings zu den Forschungsschwerpunkten des Forschungscampus und zum inhaltlichen Fokus dieser Buchreihe.

Die Buchreihe umfasst Publikationen aus den Bereichen des Engineerings, der Auslegung, Produktion und Prüfung materialhybrider Strukturen. Die Skalierbarkeit und zukünftige industrielle Großserienfähigkeit der Technologien und Methoden stehen im Vordergrund der Beiträge und sichern langfristige Fortschritte in der Fahrzeugentwicklung. Ebenfalls werden Ergebnisse und Berichte von Forschungsprojekten im Rahmen des durch das Bundesministerium für Bildung und Forschung geförderten Forschungscampus veröffentlicht und Proceedings von Fachtagungen und Konferenzen im Kontext der Open Hybrid LabFactory publiziert.

Die Bände dieser Reihe richten sich an Wissenschaftler aus der Material-, Produktions- und Mobilitätsforschung. Sie spricht Fachexperten der Branchen Technik, Anlangen- und Maschinenbau, Automobil & Fahrzeugbau sowie Werkstoffe & Werkstoffverarbeitung an. Der Leser profitiert von einem konsolidierten Angebot wissenschaftlicher Beiträge zur aktuellen Forschung zu hybriden und multifunktionalen Strukturen.

This book series presents key future topics and current work from the Open Hybrid LabFactory research campus funded by the Federal Ministry of Education and Research (BMBF) to a broad public. Discussing recent approaches and research findings based on methods and technologies for the design and large-scale production of hybrid and multifunctional structures, it highlights new production and simulation processes, as well as aspects of component functionalization and integrated life-cycle engineering.

The book series comprises publications from the fields of engineering, design, production and testing of material hybrid structures. The contributions focus on the scalability and future industrial mass production capability of the technologies and methods to ensure long-term advances in vehicle development. Furthermore, the series publishes reports on and the findings of research projects within the research campus, scientific papers as well as the proceedings of conferences in the context of the Open Hybrid LabFactory.

Intended for scientists and experts from the fields of materials, production and mobility research; technology, plant and mechanical engineering; automotive & vehicle construction; and materials & materials processing, the series showcases current research on hybrid and multifunctional structures.

Thomas Vietor
Editors

Life Cycle Design & Engineering of Lightweight Multi-Material Automotive Body Parts

Results from the BMBF sponsored collaborative research project MultiMaK2

 Springer

Editor
Thomas Vietor
Institute for Engineering Design (IK)
TU Braunschweig
Braunschweig, Niedersachsen, Germany

ISSN 2524-4787   ISSN 2524-4795   (electronic)
Zukunftstechnologien für den multifunktionalen Leichtbau
ISBN 978-3-662-65275-6   ISBN 978-3-662-65273-2   (eBook)
https://doi.org/10.1007/978-3-662-65273-2

Responsible Editor: Eric Blaschke
This Springer imprint is published by the registered company Springer-Verlag GmbH, DE, part of Springer Nature.
The registered company address is: Heidelberger Platz 3, 14197 Berlin, Germany

Research Campus Open Hybrid LabFactory

Final report of the BMBF-sponsored collaborative research project

MultiMaK2

Development of design and evaluation tools for ecologically optimized multi-material component concepts in automotive large-scale production

This research and development project was sponsored by the Federal Ministry of Education and Research (BMBF) Project within the funding initiative "Research campus—public-private partnership for innovations" and supervised by the Project Management Agency Karlsruhe (PTKA).

Funding References:

02PQ5110	Technische Universität Braunschweig
02PQ5111	IAV GmbH Ingenieurgesellschaft Auto und Verkehr
02PQ5112	ifu Institut für Umweltinformatik Hamburg GmbH
02PQ5113	INVENT Innovative Verbundwerkstoffe Realisation und Vermarktung neuer Technologien GmbH
02PQ5114	iPoint-systems gmbh
02PQ5115	Volkswagen AG

Project Period: 01.01.2015 bis 31.12.2018

SPONSORED BY THE

Federal Ministry of Education and Research

Die Dokumentation stellt die Ergebnisse aus einem Forschungs- Verbundprojekt dar. Verantwortlich für den Inhalt sind die jeweilig genannten Autoren. Ergebnisse, Meinungen und Schlüsse dieses Buches sind nicht notwendigerweise die der IAV GmbH, ifu GmbH, INVENT GmbH, iPoint gmbh, Volkswagen AG oder BASF SE.

The documentation presents the results of a collaborative research project. The respective authors are responsible for the content. The results, opinions or conclusions of this book are not necessarily those of the IAV GmbH, ifu GmbH, INVENT GmbH, iPoint gmbh, Volkswagen AG or BASF SE.

Project Consortium

Technische Universität Carolo-Wilhelmina zu Braunschweig

- Institut für Konstruktionstechnik (IK)
- Institut für Werkzeugmaschinen und Fertigungstechnik (IWF)

IAV GmbH Ingenieurgesellschaft Auto und Verkehr
ifu Institut für Umweltinformatik Hamburg GmbH
INVENT Innovative Verbundwerkstoffe Realisation und Vermarktung neuer Technologien GmbH
iPoint-systems gmbh
Volkswagen AG
BASF SE (associated)

Contents

Editors and Contributors

Makram Abdelwahed IAV GmbH, Gifhorn, Deutschland

Antal Dér Institute of Machine Tools and Production Technology (IWF), Technische Universität Braunschweig, Braunschweig, Deutschland

Tim Fröhlich Institute for Engineering Design (IK), Technische Universität Braunschweig, Braunschweig, Deutschland

Sebastian Gellrich Institute of Machine Tools and Production Technology (IWF), Technische Universität Braunschweig, Braunschweig, Deutschland

Andreas Genest Ifu Hamburg GmbH, Hamburg, Deutschland

Prof. Dr.-Ing. Christoph Herrmann Institute of Machine Tools and Production Technology (IWF), Technische Universität Braunschweig, Braunschweig, Deutschland

Andreas Kabelitz IAV GmbH, Gifhorn, Deutschland

Alexander Kaluza Institute of Machine Tools and Production Technology (IWF), Technische Universität Braunschweig, Braunschweig, Deutschland

Sebastian Kleemann Institute for Engineering Design (IK), Technische Universität Braunschweig, Braunschweig, Deutschland

Dr.-Ing. Fabian Preller INVENT GmbH, Braunschweig, Deutschland

Andreas Schiffleitner iPoint-Austria gmbh, Wien, Österreich

Christopher Schmidt Institute of Machine Tools and Production Technology (IWF), Technische Universität Braunschweig, Braunschweig, Deutschland

Lars Spresny IAV GmbH, Gifhorn, Deutschland

Tobias Steinert Ifu Hamburg GmbH, Hamburg, Deutschland

Dr.-Ing. Sebastian Thiede Institute of Machine Tools and Production Technology (IWF), Technische Universität Braunschweig, Braunschweig, Deutschland

Prof. Dr.-Ing. Thomas Vietor Institute for Engineering Design (IK), Technische Universität Braunschweig, Braunschweig, Deutschland

Introduction

1

Sebastian Kleemann, Alexander Kaluza, Tim Fröhlich, Sebastian Gellrich, Antal Dér, Thomas Vietor and Christoph Herrmann

Abstract

The present book presents selected results of the research project MultiMaK2 that has been carried out at the Research Campus Open Hybrid LabFactory in Wolfsburg, Germany. The project aimed at providing innovative engineering methods and tools that help to bring forward lightweight automotive body parts with low environmental impacts over their life cycle. The engineering of lightweight body parts is influenced

S. Kleemann · T. Fröhlich (✉) · T. Vietor
Institute for Engineering Design (IK), Technische Universität Braunschweig, Braunschweig, Deutschland
e-mail: t.froehlich@tu-braunschweig.de

S. Kleemann
e-mail: s.kleemann@tu-braunschweig.de

T. Vietor
e-mail: t.vietor@tu-braunschweig.de

A. Kaluza · S. Gellrich · A. Dér · C. Herrmann
Institute of Machine Tools and Production Technology (IWF), Technische Universität Braunschweig, Braunschweig, Deutschland
e-mail: a.kaluza@tu-braunschweig.de

S. Gellrich
e-mail: s.gellrich@tu-braunschweig.de

A. Dér
e-mail: a.der@tu-braunschweig.de
C. Herrmann e-mail: c.herrmann@tu-braunschweig.de

© Springer-Verlag GmbH Germany, part of Springer Nature 2023
T. Vietor (ed.), *Life Cycle Design & Engineering of Lightweight Multi-Material Automotive Body Parts*, Zukunftstechnologien für den multifunktionalen Leichtbau, https://doi.org/10.1007/978-3-662-65273-2_1

by innovative materials and production technologies that enable new designs. However, the indluence on resulting life cycle impacts is not transparent at engineering stages. To bridge that gap, the project promotes an integrated Life Cycle Design & Engineering when engineering lightweight automotive body parts. Therefore, the research fields of body part design, body part manufacturing as well as a concurrent life cycle engineering are introduced, and key research demands are formulated. On this basis, the subsequent chapters of this book present research results of the MultiMaK2 project.

1.1   Demand for Automotive Lightweight Body Parts

The development of new vehicle generations in automotive industry is influenced by global challenges and sectoral trends. This includes efforts to mitigate greenhouse gas emissions and other negative environmental impacts within the vehicle life cycle. One regulatory driver is the EU directive 443/2009, which limits use stage CO_2 emissions for future vehicle generations. To address these targets, a multitude of coordinated technological efforts related to the design of vehicles are required. One starting point is drivetrain innovations that lead to remarkable efficiency gains (International Council on Clean Transportation Europe 2018). However, their potential on reducing greenhouse gas emissions in vehicle use is dampened by the steady increase in vehicle curb weights. For example, the average mass of new cars in the EU increased by 10% between 2002 and 2017 (International Council on Clean Transportation Europe 2018). Those result from increased safety and comfort requirements that lead to the introduction of additional vehicle parts or an adapted dimensioning of existing parts. Another major development leading to increased curb weights is the introduction of electrified drivetrains. Due to the lower energy storage capacity per kilogramme compared to gasoline or diesel fuel tanks, longer vehicle ranges require increased vehicle weights (EEA 2018). In general, vehicle use stage fuel or energy demands increase with vehicle curb weights (Egede 2017; Trautwein et al. 2011). Fuel and electricity supply are associated with environmental impacts, e.g. related to greenhouse gas emissions per litre fuel or kilowatt hour. Therefore, heavier vehicles cause increased environmental impacts in vehicle use.

The reduction of vehicle weight through lightweight body parts is a major strategy to counteract the trade-off between increasing vehicle functionality, weight and environmental impacts. The body accounts for approximately 30% of the total weight of a passenger vehicle (Gude et al. 2015). Fig. 1.1 shows the weight proportion of the body in white to the entire vehicle and the influence of the mass on the different driving resistances. Lightweight body parts in general enable a technical functionality at lower weight than generally achievable by other means. This could encompass using less material or providing more or improved functionality per unit of weight (Klein 2013; Herrmann et al. 2018). Therefore, introducing lightweight body parts enables to improve driving dynamics. This could be leveraged to realize an improved vehicle performance or

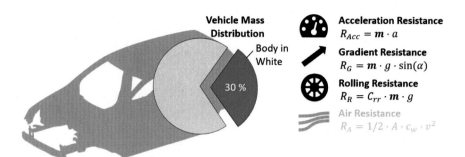

Fig. 1.1 Weight proportion of the body in white to the entire vehicle based on (Gude et al. 2018) and influence of the mass on driving resistances

downsize the drivetrain while retaining performance criteria (Alonso et al. 2012). As an observable result of light weighting efforts in automotive industry, curb weights of conventional vehicles remained constant or even decreased for new vehicle generations within recent years (International Council on Clean Transportation Europe 2018; Lehmhus et al. 2015).

Anticipating that there will be stricter targets for environmental impacts in vehicle use stage as well as increasing requirements on vehicle safety and comfort, there is an ongoing need for vehicle light weighting. While there are several strategies to realize lightweight body parts, material substitution dominates past efforts in automotive industry (Gude et al. 2018). However, large-scale manufacturing of lightweight body parts is subject to economic constraints. Long manufacturing cycle times, less mature manufacturing technologies, energy- and cost-intensive raw materials as well as large investment needs inhibit an extensive switch from steel structures to aluminium, magnesium or composites (Gude et al. 2018). Fig. 1.2 illustrates strategies for large-scale production of lightweight body parts and place these to an expected time horizon. Today, automotive manufacturing only includes small quantities of fibre-reinforced plastics (FRP); aluminium body parts are produced in medium quantities and steel designs in large quantities. Medium-term developments are expected to enable larger-scale productions for FRP and mixed designs between aluminium and steel. In perspective, multi-material design offers a promising approach for the large-scale production of lightweight body parts by combining metals, plastics and fibre reinforcements on a part level. In line with that development, design methods need to be enhanced to fully exploit the potential of different materials.

1.2 Engineering of Automotive Lightweight Body Parts—the MultiMaK2 Approach

The present book compiles methods and tools developed within the project MultiMaK2 sponsored by the German Federal Ministry of Education and Research (BMBF) within the Forschungscampus Open Hybrid LabFactory and managed by the Project

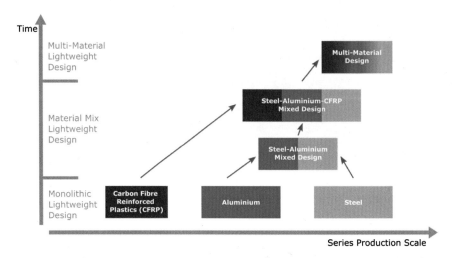

Fig. 1.2 Strategies for large-scale production of lightweight body parts, adapted from Goede et al. (2009)

Management Agency Karlsruhe (PTKA) between January 2015 and December 2018. The project MultiMaK2 aimed at developing design and evaluation tools for lightweight multi-material vehicle body parts with low life cycle environmental impacts. An interdisciplinary approach has been followed to enable a concurrent Life Cycle Design & Engineering for multi-material lightweight body parts. This covers design engineering, manufacturing engineering as well as life cycle engineering (LCE) disciplines (see Fig. 1.3). While body part design brings forward innovative concepts, their manufacturability needs to be ensured by capabilities of respective manufacturing processes, e.g. regarding quality, cost and cycle times. On the other hand, innovative manufacturing processes can enable new designs, e.g. through integrated processing of materials from different material families. Both of the disciplines show a close relation to LCE activities. One major task is to enable a cradle to grave perspective for assessing cost and environmental impacts of the newly proposed body part designs. From a gate-to-gate perspective, LCE serves as a guidance in the development and implementation of innovative manufacturing processes with low environmental impacts, e.g. resulting from energy and material demands of manufacturing processes and their surrounding factory infrastructure.

The MultiMaK2 project brought forward methods and engineering tools addressing specific challenges of the single discipline as well as the interplay between them. The application is oriented on demands of the large-scale automotive manufacturing industry that shows a high degree of labour division. To gain insights on specific challenges for multi-material lightweight body parts for passenger cars, an exemplary design process based on the results of the European research Project SuperLIGHT-CAR has been carried out (Goede et al. 2009). Another major topic within the project

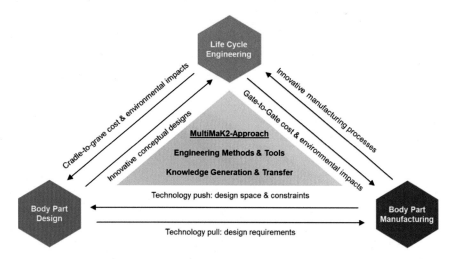

Fig. 1.3 MultiMaK2 approach at the interface between body part design, body part manufacturing and life cycle engineering of multi-material lightweight body parts, adapted from Kaluza et al. (2017)

MultiMaK2 was to transfer knowledge between design and manufacturing engineers and life cycle engineering. Visual analytics builds a methodological foundation on this matter. The tight integration of data, engineering models and results visualization towards an informed knowledge building has been elaborated both theoretically and through specific prototypes (Kaluza et al. 2018). The research infrastructure "Life Cycle Design & Engineering Lab" was established at Open Hybrid LabFactory as a platform to transfer the results obtained within MultiMaK2.

Towards achieving the project goals, several subordinate research activities have been covered:

- Identification of vehicle body parts that are well-suited for introducing multi-material designs (Sect. 1.3)
- Conceptual design of competing multi-material body part concepts and concurrent evaluation (Chap. 2)
- Knowledge management for designing multi-material lightweight body parts in automotive applications (Chap. 3)
- Derivation of quality and efficiency indicators for large-scale manufacturing scenarios based on energy and process data in prototype manufacturing (Chap. 4)
- Manufacturing planning for innovative process chains for multi-material lightweight body parts (Chap. 5)
- Life cycle assessment of multi-material lightweight body parts at early design stages, incorporating scenarios for foreground and background systems (Chap. 6)
- Visual analytics-based approaches to enable knowledge building between design, manufacturing and life cycle engineering (Chap. 7)

In the following, specific challenges and methodological approaches of the three disciplines, as well as interfaces to other disciplines, are elaborated.

1.2.1 Body Part Design

Engineering methods and tools in body part design require to meet adapted specifications when considering multi-material lightweight designs. This especially originates from a broad solution space that opens up when combining different materials on a part level (see Fig. 1.4). In contrast to mono-material designs, not only part geometry and material properties are influencing the final design, but also different materials can be combined within the installation space of a specific body part. For example, body parts might allow to realize a similar mechanical performance by combining steel and local FRP reinforcements or by manufacturing a body part from aluminium. Within the multi-material design, alloy compositions could be varied for the metallic sub-part, while mechanical performance of the fibre reinforcement is subject to fibre properties, e.g. strength, fibre orientation or fibre content. Therefore, adapted engineering methods are required to assist the conceptual design and identification of preferred design options. The MultiMaK2 project targets methods and tools that enable to handle the increased solution space. This includes an identification of body parts with high potential to be realized in multi-material design (Sect. 1.3). Further, Chap. 2 introduces a development procedure that considers specific demand-oriented product models on different level of abstraction for the body part concepts shown in. Both lead to numerous different multi-material concepts.

In order to describe the complexity of multi-material design in more detail, a classification based on the physical behaviour of materials is followed and illustrated in Fig. 1.5 (Nestler 2014; Kleemann et al. 2017). By combining different materials within a single

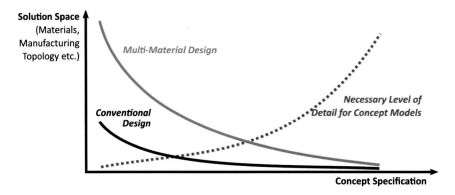

Fig. 1.4 Solution space for hybrid lightweight body parts within automotive development with demand for situation-oriented concept models

body part, the complexity regarding the part's design and manufacturing is increasing due to dissimilar material properties. At the lowest level of multi-material design, the combined materials belong to the same type of material. Automotive bodies made of different steel grades, for example, have been state of the art since the 1990s. On the 2nd level, materials from the same main group are combined, for example different metals. As a combination of steel and aluminium, this mixed design has been used in vehicle bodies since the 2000s, for example in the Mercedes C-Class (Maurer et al. 2014) and the Audi TT (Seehafer 2014). Another example for 2nd level multi-material design are structures made of composites and thermoplastics, for instance composite profiles with injection-moulded ribs. On the 1st level, all materials can be combined with each other. This design approach is the focus of numerous research projects and centres such as the Open Hybrid LabFactory (Fischer et al. 2014) and SMiLE (Kothmann et al. 2018).

A major challenge in developing multi-material designs is limited experience and representations of formalized knowledge within automotive engineering. This includes knowledge about individual materials, such as FRP, as well as applicable production and joining technologies. Especially in case of composites, this knowledge is already available in other industries, such as aerospace. However, the complexity of aerospace design, manufacturing and quality management is not compatible with the requirements of the automotive industry and thus not directly transferable. Within MultiMaK2, a knowledge management database was developed to support in developing multi-material body parts (Chap. 3). The knowledge was derived from literature and existing multi-material-designs as well as from the findings of the concept development in particular.

1.2.2 Body Part Manufacturing

Established manufacturing processes for automotive body parts are based on sheet metal processing (Ingarao et al. 2011). Fibre-reinforced plastics (FRP)-based multi-material

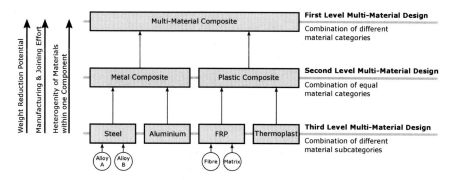

Fig. 1.5 Level of different multi-material designs based on (Kleemann et al. 2017)

structures, however, exploit their full lightweight potential in load-path optimized structures, which calls for the development of new manufacturing processes and process chains that fulfil the requirements of high-volume production (Buschhoff et al. 2016). Fig. 1.6 illustrates an overview of a process chain combining metals and FRP for manufacturing automotive structural body parts. FRP process chains encompass textile processes for semi-finished parts processing and a number of processes depending on part geometry and matrix material for final part production (Dröder et al. 2014). Hybrid process chains emerge from the combination of intrinsic metal and FRP processing and show promising results towards being a competitive alternative to traditional parts manufacturing (Fleischer et al. 2018).

Fig. 1.6 as well displays system boundaries for evaluating environmental impacts of manufacturing processes. Therefore, differentiations are made between the gate-to-gate perspective for semi-finished as well as for final parts and the cradle-to-gate perspective that encompasses the whole process chain starting from raw materials over semi-finished and final parts manufacturing. The baseline for evaluating environmental impacts in manufacturing is an inventory of energy and materials flows associated with the production of the parts. Beyond process-related energy demands, also the energy demands arising from the technical building services of the factory environment need to be considered. With this regard, a simulation-based approach for the estimation of gate-to-gate energy demands for new manufacturing processes for multi-material lightweight body parts is introduced (Chap. 5). Due to the presence of new or adapted process technologies for the manufacturing of high-volume multi-material lightweight body parts, a lack

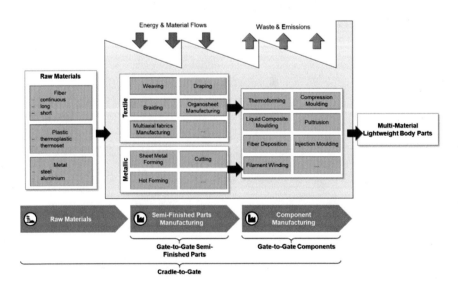

Fig. 1.6 Environmental evaluation perspectives in manufacturing process chains for multi-material lightweight body parts (manufacturing processes based on (Fleischer et al. 2018; Dröder et al. 2014))

of reliable life cycle inventory data to build quantitative models exists (Herrmann et al. 2018). Data-based methods, in terms of cyber physical production systems CPPS), can serve as a potential approach for the acquisition of reliable manufacturing data. Those methods can aid in simulation parameterization by reducing the design freedom to robust and efficient process settings. Moreover, an automated deviation of part-specific energy demand can be enabled through a data-based machine state recognition. These and additional levers of cyber physical production systems for the manufacturing of multi-material lightweight body parts are discussed in Chap. 4.

1.2.3 Life Cycle Engineering

Within large-scale automotive manufacturers, there is a consensus to develop future vehicle generations with the goal to decrease life cycle environmental impacts, as exemplary described by (Broch et al. 2015). Engineering methods and tools that are implemented to serve that goal are subsumed as "Life Cycle Engineering (LCE)". Since its beginnings, LCE is defined as designing the "[…] product life cycle through choices about product concepts, structure, materials and processes […]" with life cycle assessment (LCA) being "[…] the tool that visualizes the environmental and resource consequences of these choices" (Alting 1995).

Multi-material alternatives should be designed to carry lower environmental loads than reference designs over the entire life cycle, including raw materials provision, manufacturing, use and end-of-life (Herrmann et al. 2018). Fig. 1.7 qualitatively shows potential environmental impacts alongside a full vehicle life cycle considering different lightweight material options. From a perspective of greenhouse gas emissions, cradle-to-gate impacts from raw materials provision of common lightweight materials, namely

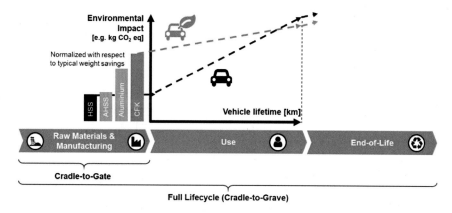

Fig. 1.7 Environmental break-even for lightweight body part concepts in relation to conventional designs (qualitative representation), based on (Herrmann et al. 2018)

aluminium, magnesium and composites, are likely to exceed impacts from conventional steel materials per kg material (Herrmann et al. 2018). As previously described, gate-to-gate manufacturing processes for lightweight materials might be less efficient compared to steel processing due to their potential lower technical maturity. Similar observations are made for end-of-life processing (Herrmann et al. 2018). In addition, conventional vehicle structures undergo established end-of-life treatments with secondary materials made available to material markets. Lightweight body parts, especially when combining different materials with new joining technologies, might lead to decreasing amounts of secondary materials to be reintroduced into the markets (Soo 2018).

Within MultiMaK2, LCE was applied to analyse environmental burdens and trade-offs between different life cycle stages and impact categories at an early stage of conceptual design of multi-material body parts. Different engineering paths could be followed to identify designs that carry reduced environmental burdens over their life cycle. First, additional burdens from raw materials provision, manufacturing & end-of-life can be avoided, e.g. through optimized designs, sourcing of recycled materials or more efficient manufacturing and end-of-life processes. Second, lightweight body parts can be applied to scenarios with a high benefit within the use stage. This entails markets with comparably high environmental impacts per driven kilometre, e.g. through a high share of fossil fuels or vehicle applications with high mileages during lifetime. Third, the functionality of lightweight body parts can be extended. This entails the integration of a set of parts instead of a one-to-one substitution, e.g. by integrating further functions like thermal insulation (Herrmann et al. 2018; Kaluza et al. 2017; Broch et al. 2015).

Chap. 6 discusses environmental impacts of multi-material lightweight body parts covering different scenarios. This information can be used as a direct feedback to engineering activities, e.g. dimensioning of a body part or the planning of a manufacturing process chain. On the other hand, the influence of renewable electricity shares on the product system is studied.

1.3   Reference Body Parts of the Project

Legal requirements regarding greenhouse gas emission of vehicles and the growing customers' demand of environmental friendly mobility increase the need for lightweight design in automotive industry. However, there are many other development goals to consider in addition to the reduction of weight. Examples are the product's resulting costs or its manufacturability. Furthermore, the environmental impact of a vehicle cannot only be determined by the fuel consumption, but requires a consideration of the entire product life cycle. Consequently, the specific opportunities depend on the reference part to be replaced as well as the basic scenario of development, manufacturing and application. In relation to a steel part, multi-material design offers the potential of weight reduction as well as a reduction of manufacturing costs, as many time-intensive manufacturing and joining steps become no longer necessary. On the contrary, in case of a CFRP-intensive

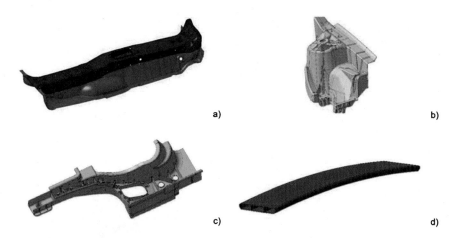

Fig. 1.8 Identified body parts for application of multi-material design **a** centre tunnel, **b** suspension strut mounting, **c** casting rail, **d** roof reinforcement (Volkswagen, 2019)

reference, multi-material design offers the potential to reduce both life cycle costs and environmental impact. Therefore, the identification of different reference body parts is necessary to analyse the broad potentials of multi-material design entirely.

The superordinate reference for MultiMaK2 is given by the SuperLIGHT-CAR (SLC) as presented by (Goede et al. 2009). Acting as a reference for the development of multi-material parts in Chap. 2, four body parts were identified as described by Volkswagen (2019) and shown in Fig. 1.8.

The parts show very different specific geometries, installation conditions, load applications, applied materials and manufacturing processes. In addition, the parts' property values offer very different scenarios for the development of concepts. For example, the centre tunnel out of steel offers high potential regarding a reduction of weight, whereas the roof reinforcement out of carbon fibre-reinforced plastic (CFRP) offers high potential of reducing the environmental impact. The identified reference body parts allows the consideration of very different multi-material design approaches and, therefore, offer good examples for the application of the design and evaluation tools to be developed within the project MultiMaK2.

1.4   Contents and Structure of the Book

The present book compiles engineering methods and implemented tools within the project MultiMaK2. The chapters are organized based on the three engineering disciplines as shown in Fig. 1.9. Following the introduction, this includes seven chapters that discuss specific challenges and show methodical paths that have been followed during the project. All methods and tools have been implemented within the Life Cycle Design & Engineering Lab at Open Hybrid LabFactory (Chap. 7).

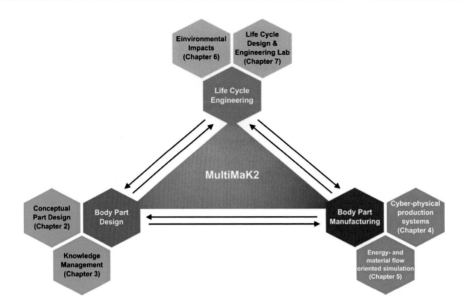

Fig. 1.9 Contents and structure of the book

References

Alonso, E., Lee, T. M., Bjelkengren, C., Roth, R., & Kirchain, R. E. (2012). Evaluating the potential for secondary mass savings in vehicle lightweighting. Environmental Science and Technology, 46(5), 2893–2901. doi: https://doi.org/10.1021/es202938m

Alting, L. (1995). Life Cycle Engineering and Design. CIRP Annals - Manufacturing Technology, 44(2), 569–580. doi: https://doi.org/10.1016/S0007-8506(07)60504-6

Broch, F., Warsen, J., & Krinke, S. (2015). Implementing Life Cycle Engineering in Automotive Development as a Helpful Management Tool to Support Design for Environment. In G. Sonnemann & M. Margni (Eds.), Life Cycle Management (pp. 319–329). Springer. doi: https://doi.org/10.1007/978-94-017-7221-1

Buschhoff, Clemens; Brecher, C.; Emonts, M. (2016): High volume production of lightweight automotive structures. In: Michael Bargende, Hans-Christian Reuss und Jochen Wiedemann (Hg.): 16. Internationales Stuttgarter Symposium. Wiesbaden: Springer Fachmedien Wiesbaden, S. 213–226. DOI https://doi.org/10.1007/978-3-658-13255-2

Dröder, K.; Herrmann, C.; Raatz, A.; Große, T.; Schönemann, M.; Löchte, C. (2014): Symbiosis of plastics and metals: integrated manufacturing of functional lightweight structures in high-volume production. In: Kunststoffe im Automobil-bau. Mannheim, S. 31–44.

EEA. (2018). Electric vehicles from life cycle and circular economy perspectives - TERM 2018: Transport and Environment Reporting Mechanism (TERM) report. doi: https://doi.org/10.2800/77428

Egede, P. (2017). Environmental Assessment of Lightweight Electric Vehicles. Cham: Springer International Publishing. doi: https://doi.org/10.1007/978-3-319-40277-2

Fischer, F.; Kleemann, S.; Vietor, T. (2014): Smart Production of Hybrid Material Automotive Structures at ForschungsCampus Wolfsburg in the "Open Hybrid LabFactory". In: Conference proceedings - ITHEC 2014, 2nd International Conference & Exhibition on Thermoplastic Composites.

Fleischer, J.; Teti, R.; Lanza, G.; Mativenga, P.; Möhring, H-C.; Caggiano, A. (2018): Composite materials parts manu-facturing. In: CIRP Annals 67 (2), S. 603–626. DOI: https://doi.org/10.1016/j.cirp.2018.05.005.

Goede, M., Stehlin, M., Rafflenbeul, L., Kopp, G., & Beeh, E. (2009). Super Light Car-lightweight construction thanks to a multi-material design and function integration. European Transport Research Review, 1(1), 5–10. doi: https://doi.org/10.1007/s12544-008-0001-2

Gude, M., Lieberwirth, H., Meschut, G., & Zäh, M. F. (2015). FOREL-Studie - Chancen und Herausforderungen im ressourceneffizienten Leichtbau für die Elektromobilität. Retrieved from http://plattform-forel.de/ (last visited 06.03.2020)

Gude, M., Lieberwirth, H., Merschut, G., Tekkaya, A. E., & Zäh, M. F. (2018). FOREL Studie 2018. Retrieved from https://plattform-forel.de/studie/ (last visited 06.03.2020)

Herrmann, C., Dewulf, W., Hauschild, M., Kaluza, A., Kara, S., Skerlos, S., & Engineering, L. C. (2018). Life cycle engineering of lightweight structures. CIRP Annals, 67(2), 651–672. doi: https://doi.org/10.1016/j.cirp.2018.05.008

Ingarao, G.; Di Lorenzo, R.; Micari, F. (2011): Sustainability issues in sheet met-al forming processes: an overview. In: Journal of Cleaner Production 19 (4), S. 337–347. DOI: https://doi.org/10.1016/j.jclepro.2010.10.005.

International Council on Clean Transportation Europe. (2018). EUROPEAN VEHICLE MARKET STATISTICS Pocketbook 2018/2019. Retrieved from https://www.theicct.org/sites/default/files/publications/ICCT_Pocketbook_2018_Final_20181205.pdf (last visited 06.03.2020)

Kaluza, A., Kleemann, S., Fröhlich, T., Herrmann, C., & Vietor, T. (2017). Concurrent Design & Life Cycle Engineering in Automotive Lightweight Component Development. Procedia CIRP, 66, 16–21. doi: https://doi.org/10.1016/j.procir.2017.03.293

Kaluza, A., Gellrich, S., Cerdas, F., Thiede, S., & Herrmann, C. (2018). Life Cycle Engineering Based on Visual Analytics. Procedia CIRP, 69, 37–42. doi: https://doi.org/10.1016/j.procir.2017.11.128

Kleemann, S., Inkermann, D., Bader, B., Türck, E., & Vietor, T. (2017). A Semi-Formal Approach to Structure and Access Knowledge for Multi-Material-Design. In 21st International Conference on Engineering Design (ICED17). doi: https://doi.org/10.24355/dbbs.084-201708301114

Klein, B. (2013). Leichtbau-Konstruktion. Wiesbaden: Springer Fachmedien Wiesbaden. doi: https://doi.org/10.1007/978-3-658-02272-3

Kothmann, M.H., Hillebrand, A. & Deinzer, G. (2018): Multi-material bodies for battery-electric vehicles. Lightweight des worldw 11, 6–13. DOI: https://doi.org/10.1007/s41777-018-0005-0.

Lehmhus, D., von Hehl, A., Kayvantash, K., Gradinger, R., Becker, T., Schimanski, K., & Avalle, M. (2015). Taking a downward turn on the weight spiral - Lightweight materials in transport applications. Materials and Design, 66(PB), 385–389. doi: https://doi.org/10.1016/j.matdes.2014.10.001

Maurer, S., Wittner, B., Sikorski, S., Fuchs, D., Brand, D., & Hug, T. (2014): Der Wunsch nach weniger. ATZextra, 19(8), 62–67. DOI: https://doi.org/10.1365/s35778-014-1221-9.

Volkswagen AG: MultiMaK2 - Entwicklung von Design- Und Bewertungstools Für Nutzungsgerecht Ökologisch Optimierte Multi-Material-KFZ-Bauteilkonzepte in Der Großserie, Teilprojekt: Bauteilidentifikation, LCA Fahrzeug-Nutzungsszenarien, Handbuch Multi-Materialbauweise : Schlussbericht : Förderprogramm: Förderinitiative Forschungscampus - Öffentlich-Private Partnerschaft Für Innovationen, Forschungscampus Open Hybrid LabFactory : Laufzeit Des Vorhabens von: 01.01.2015 Bis: 31.12.2018. [Volkswagen AG], 2019, doi: https://doi.org/10.2314/KXP:1693272911

Nestler, D. J. (2014). Beitrag zum Thema VERBUNDWERKSTOFFE - WERKSTOFFVERBUNDE.

Seehafer, G. (2014): Karosseriestruktur des neuen Audi TT. ATZ Automobiltech Z 116, 56–61. DOI: https://doi.org/10.1007/s35148-014-0451-x.

Soo, V. K. (2018). Life Cycle Impact of Different Joining Decisions on Vehicle Recycling. doi: https://doi.org/10.25911/5d6907c5dd810.

Trautwein, T., Henn, S., & Rother, K. (2011). Weight Spiral adjusting Lever in Vehicle Engineering. ATZ worldwide eMagazine, 113(5), 30–35. doi: https://doi.org/10.1365/s38311-011-0053-0

Development of Automotive Body Parts in Multi-Material Design—Processes and Tools

Tim Fröhlich, Sebastian Kleemann, Thomas Vietor, Lars Spresny, Makram Abdelwahed, Andreas Kabelitz and Fabian Preller

Abstract

Besides the various opportunities of multi-material design, such as weight reduction or function integration, additional challenges occur within the design process. On the one hand, the rising complexity of multi-material body parts requires an additional assistance for the designer. On the other hand, it is of great importance to estimate the developed concepts' properties—environmental properties in particular—at a very

T. Fröhlich (✉) · S. Kleemann · T. Vietor
Institute for Engineering Design (IK), Technische Universität Braunschweig, Braunschweig, Deutschland
e-mail: t.froehlich@tu-braunschweig.de

S. Kleemann
e-mail: s.kleemann@tu-braunschweig.de

T. Vietor
e-mail: t.vietor@tu-braunschweig.de

L. Spresny · M. Abdelwahed · A. Kabelitz
IAV GmbH, Gifhorn, Deutschland
e-mail: lars.spresny@iav.de

M. Abdelwahed
e-mail: makram.abdelwahed@iav.de

A. Kabelitz
e-mail: andreas.kabelitz@iav.de

F. Preller
INVENT GmbH, Braunschweig, Deutschland
e-mail: Fabian.Preller@invent-gmbh.de

© Springer-Verlag GmbH Germany, part of Springer Nature 2023
T. Vietor (ed.), *Life Cycle Design & Engineering of Lightweight Multi-Material Automotive Body Parts*, Zukunftstechnologien für den multifunktionalen Leichtbau, https://doi.org/10.1007/978-3-662-65273-2_2

early stage of development in order to focus on promising concepts. To solve these challenges, this chapter introduces a procedure for the development of multi-material body parts on different levels of abstraction. It aims at reducing the design task's complexity already at a very early stage of design. Among all considered properties, the focus is especially on the environmental properties over the entire life cycle. The procedure is applied on two body parts of the SuperLIGHT-CAR in a case study in order to develop multi-material body parts and validate the procedure as well as the additional tools developed within the project.

2.1 Introduction

Multi-material design offers a high degree of freedom in designing automotive body parts due to different design approaches or material combinations and, therefore, adapt the body part to the specific requirements and boundary conditions. On the contrary, this causes a more complex design task due to the wide solution space. Common numerical simulations especially for crash analyses are very time intensive regarding execution as well as pre- and postprocessing. In addition, many development goals (e.g. weight, costs or environmental impact) and boundary conditions (e.g. restricted installation spaces) must be considered. In order to exploit respectively systematically reduce the solution space, these requirements must be considered as early as possible in concept development. Otherwise, the most suitable design option might not be considered at all or unsuitable concepts are designed in detail although they did not fit the requirements from the beginning. As a result, the designer needs an assistance in analysing different design options to find those that are able to fulfil the development goals best possible with an appropriate effort. These aspects reveal the need for a stepwise multi-material concept development with increasing level of detail and an opportunity for concurrent engineering.

Therefore, specific information on all different level of concept specification must be available.

Within the project MultiMaK2, a procedure was developed that represents the body part concepts by different product models on different level of abstraction from an analytical description to FE simulations on body parts and entire vehicle level. By this development procedure, multi-material concepts are developed as replacement for predecessor body part from the SuperLIGHT-CAR (SLC) (Goede et al. 2009). In this chapter, the concept development is shown exemplary for a roof reinforcement as well as for a centre tunnel, as identified in Sect. 1.3. However, the concept development within the project does not predominantly aim at finding the optimum solution for the multi-material body part and detailing it to series maturity. It aims at developing different concept variants for different scenarios in order to show the complexity of opportunities of multi-material design. Furthermore, the concepts serve as use cases for testing the procedure and the tools developed within the project. The main focus is on which information can be derived or estimated by the different level of abstraction and, basing on this, which decision can be made regarding concept selection or rejection.

2.2 Review on Development and Lightweight Design Approaches

The state of the art offers a high potential in adapting existing approaches regarding development processes and goals as well as material selection, lightweight design in general and multi-material design in particular. This section gives a short review on the considered approaches.

2.2.1 Development Processes and Goals

The development of automotive body parts in series production usually follows a standardized process. The individual development goals of the body parts are determined by a breakdown of the entire vehicle over assemblies to the individual body parts in a very early development planning stage (Weber 2009; Kaluza et al. 2016). The validation and verification, again, is done by analysing the superordinate system.

This procedure leads to a fixed installation space and defined development goals before the actual concept development starts. The results are limited opportunities in designing the body part especially by alternative lightweight materials or even material combination, such as by multi-material design. For example, aluminium requires an extended installation space in order to exploit its full potential. CFRP has great potential to reduce a body parts weight by its high specific stiffness and strength. However, the costs and environmental impact might exceed the development goals (Duflou et al. 2009; Song et al. 2009). A discrepancy of these development goals can only be compensated with difficulty by other components if the goals and installation spaces are already defined.

As a result, the potentials of different design approaches, including different materials and their combinations, must be considered in parallel to the definition of development goals. This can help to achieve the development goals as best as possible, especially with regard to resolving goal conflicts, and to reduce compromises. Due to the fundamentally different behaviour of the possible design opportunities and a possible missing expert knowledge regarding all opportunities, a fundamental analysis of these resulting properties in an early development phase seems to be useful.

However, these properties (such as weight or costs) can only be defined indirectly by the definition of the related characteristics (such as by materials or dimensions). Therefore, a structured analysis of the relations between characteristics and properties is necessary in order to identify the possible scope of influence by the designer. An approach to structure these relations is given by the Property Driven Development (CPM/PDD) (Weber 2007). (Köhler et al. 2008) again extended this approach by a matrix representation, which maps all the relations between characteristics and properties. Both approaches can be implemented in an analytical definition of the body part to be developed and, therefore, offer high potential for an early property estimation of multi-material concepts.

2.2.2 Material Selection and Lightweight Design Approaches

Material selection is a major task in the design process of automotive body parts, as the demanded properties and functions can only be realized by the application of material to a specific geometry or vice versa. For material selection, there are many approaches existing taking into account different development goals already on a very abstracted level of concept specification. Examples are, material indices, Ashby charts (Ashby 2005) or key performance indicators (Klein 2013). An overview of approaches for material selection approaches is given by Ashby, Jahan et al. or Babanli et al. considering different application scenarios (Ashby et al. 2004; Jahan et al. 2010; Babanli et al. 2019).

For selecting materials, mostly a limited amount of properties, respectively evaluation criteria, are taken into account concurrently. However, due to the amount of different development goals, a multi-criteria decision-making is necessary to identify the most suitable material, such as shown by (Athawale et al. 2011). In addition, by the application of multi-material design there are more than just a single material to be identified. Then again, these materials cannot be considered individually, but only in combination taking into account their topology within the concept. This makes material selection even more complex. Examples for a material selection for multi-material design in particular are given by (Giaccobi et al. 2010) or (Sakundarini et al. 2013).

In addition to the materials themselves and the development goals to be considered, the body part's geometry has to be taken into account. Important aspects, for example, are a limited installation space or specific geometrical requirements given by the assembly process or adjacent parts.

An approach that takes into account limitations of the installation space is shown by Wanner on abstracted geometries (Wanner 2010). An example for the importance of the installation space is given in Fig. 2.1. In this example, a hollow profile from steel is replaced by a hollow profile from aluminium. In order to achieve the same bending stiffness, more material is necessary. If the outer installation space cannot be exceeded, the wall thickness can only be increased towards the neutral fibre of the profile. The result is a higher weight compared to a steel profile. Only if the installation space can be exceeded, the aluminium is able to exploit its potential.

However, the specific geometry respectively cross section of the concept to be developed can differ from the reference due to the application of different design approaches or materials. Therefore, materials and geometries have to be identified simultaneously in order to consider the dependencies between both. Consequently, an

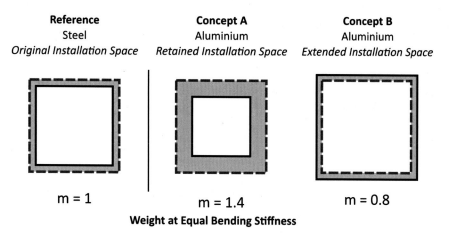

Fig. 2.1 Relative mass of a reference steel profile (**1**) and aluminium profiles with retained (**2**) and extended installation space (**3**) at same bending stiffness also considering buckling resistance

early decision-making of multi-material concepts requires an adaption of the analytical approach to compare different design approaches and estimate their properties considering multiple development goals.

For the application of lightweight design or multi-material design in particular for automotive body parts, a number of different examples is available. Kellner, for example, compared different multi-material design approaches on a generic profile level including open and closed profiles in mono- and multi-material design. Basing on the analysation results (with focus on lightweight and economical aspects) of these generic profiles, Kellner transferred suitable design approaches to complex body parts. By this, it was shown that an analysis of generic concepts can successfully be transferred to different body parts and, therefore, increase the efficiency of part design. (Kellner 2014).

Examples for automotive body parts in multi-material design are given, for example, by a FRP-steel centre tunnel from project LehoMit-Hybrid (Kuhn et al. 2019), a multi-material function integrated battery tray (Dröder 2020) or a "3D Hybrid" b-pillar (Götz et al. 2018). An overview of different examples for multi-material components from industry and research is shown by Bader et al. (Bader et al. 2019).

Within the project SuperLIGHT-CAR, an entire body in white was redesigned with focus on lightweight design including three variants with different degrees of lightweight design. Within the body, each part is only out of a single material; however, by application of different materials to the different body parts and their combination, a multi-material body was developed. The result also includes a methodology for the allocation of specific materials to different body parts. (Goede et al. 2009).

2.3   Developing Approach for Multi-Material Automotive Body Parts

General state-of-the-art development processes offer high potential to be adapted or extended in order to develop multi-material automotive body parts on different levels of abstraction. Within the project MultiMaK2, a procedure was developed in order to create multi-material concepts more efficiently. The different resulting multi-material concepts serve on the one hand as use cases for the validation the tools to be developed within this project. On the other hand, multi-material-specific design rules were derived from the concepts (Chap. 3).

The basis of the development procedure is the Extended Mapping Matrix approach by Köhler (Köhler et al. 2008) basing on the CPM/PDD approach by Weber (Weber 2007) as shown by Kleemann et al. (Kleemann et al. 2017). After identifying body parts with a potential for multi-material design (as shown in Sect. 1.3), the procedure's steps are shown in Fig. 2.2.

With respect to the complexity of the design task and the various aspects that have to be considered, the beforehand shown approach is adapted to a stepwise procedure with an increasing level of detail of the applied product models as shown in Fig. 2.3.

Level 0 considers the body part's requirements and boundary conditions. It aims at identifying the actually demanded values for the requirements rather than just transferring the properties of the predecessor part. On level 1, the first concepts are developed by choosing suitable geometries and materials that meet the requirements derived

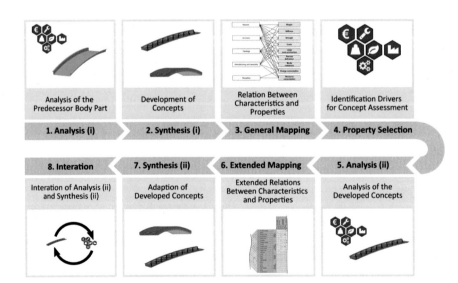

Fig. 2.2 Steps of the methodological approach towards multi-material design of automotive body parts

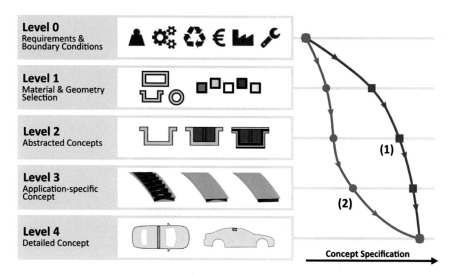

Fig. 2.3 Concept specification on different level of detail for simple body parts (**1**) and complex body parts (**2**)

before. Both are combined to general design approaches that are feasible (e.g. due to available opportunities of manufacturing). Level 2 uses a two-dimensional analytical representation of the concept. Therefore, the concept can be defined by combining different elements. On level 3, the body part is designed in CAD and finally analysed by a FE-simulation. Level 4 again considers the behaviour inside the entire vehicle respectively its influence on the entire vehicle.

The first level considers product models with a high degree of abstraction. Therefore, the gained specification of the concept highly depends on the complexity of the body part itself as well as boundary conditions. For a simple profile with only static load cases, as an example, most of the concept's specification can already be determined on level 1 or 2. The applied models are able to sufficiently describe the part's shape, material model and load cases. This is why an analytical calculation is still widely used in aviation, as many elements of, e.g., the fuselage can be described by an analytical model. In contrast, for complex body parts an analytical model might not be able to represent the body part's geometry entirely or occurring local effects. In this case, a CAD-modelling and FE-simulation (level 3) might be necessary. For example, considering complex crash load cases, an analysis of the entire vehicle is necessary. These examples show that the gained knowledge can be very different depending on the body part to be developed and its specific boundary conditions. However, there is always information that can be derived from the early analyses in order to reduce the solution space.

Subsequently, the different levels are described in further detail. Therefore, the focus is on the level 0 to 2, as levels 3 and 4 are common praxis in industry.

2.3.1 Level 0: Requirements Model

The aim of level 0 is to prepare the actual concept development in case of the body part's requirements and boundary conditions as well as the deduction of the main drivers for decision-making. In addition, the levels are defined; this decision can be made on.

At first, the development goals and boundary conditions of the part have to be elaborated, which lead to the requirements for concept development. This can be done by simply analysing the predecessor's properties. However, even if the concept to be developed directly substitutes the predecessor part, it is of high importance not only to transfer the predecessor's properties as requirements for the new concept. As some of the body part's properties are dependent (as mentioned in the basic approach), some values might exceed the actual requirements due to the applied design approach. Therefore, it is of high importance that the really demanded requirements and boundary conditions rather than only the properties of the predecessor parts are identified. This includes alterations of the given boundary conditions. For example, the opportunity to extend the body parts installation space when applying lightweight materials can lead to a huge decrease of weight.

As the amount of different requirements is difficult to handle during conceptualization, the main drivers for decision-making are identified. Depending of the specific application, these drivers can be very different. On the one hand, they can depend on the specific focus of the design task (e.g. weight reduction, cost reduction, GWP reduction or increasing of mechanical performance). On the other hand, a crash-relevant body part has to be analysed by a crash simulation of the entire vehicle. Therefore, a planning, on which level the information can be gained, is necessary. Furthermore, if the necessary property cannot be estimated directly, analogous models have to be identified. For example, the behaviour for stiffness and strength of a simple profile out of a single material with clear load cases (such as bending) usually can sufficiently be described by level 1. For a hybrid profile under the same conditions, level 2 might be necessary. However, a body part with a complex shape can only sufficiently described by a more detailed geometry on level 3. In case of crash loads, it has to be analysed on level 4 taking into account the entire vehicle in order to determine the body part's properties.

In case of the manufacturing, the given opportunities have to be identified. This aspect highly depends on the existing manufacturing technologies and the need for a large-scale production. In addition, the assembly in the body in white gives decent boundary conditions to the concepts.

2.3.2 Level 1: Material, Geometry and Design Model

Level 1 is the start of conceptualization for the multi-material body part. In this step, the materials and the part's basic geometries are defined. This leads to the definition of the general design approach as well. A first, decision-making can be done by considering material properties in combination with the concept's general geometry. By this, a first estimation of the concept's properties can be made, e.g. mechanical behaviour, weight, costs or environmental impact. The consideration of more specific design approaches can already include certain aspects of manufacturing and assembly.

In order to develop first design approaches for the multi-material concept, suitable materials and geometries have to be identified regarding the specific requirements. As this level is very generic, it can only provide a first impression of the corresponding properties. However, unsuitable materials or geometries can be rejected within this step in order to reduce complexity of the development task.

Since the demanded functions of a body part can only be realized by the application of material to a geometry, the selection of the general geometry is a major task in part design as well. If there is a general freedom in design, it can be reasonable to consider different geometries, respectively cross sections. Besides the resulting mechanical properties the manufacturing, an accessibility for further assembly of parts is main driver for geometry selection. All these aspects can already be taken into account within this level.

The main factor in designing a body part is its mechanical behaviour, which defines, e.g., wall thicknesses and, therefore, other properties, such as costs, weight or environmental impact. A way to identify these properties is a comparison of different materials by specific material indices (Ashby 2005; Klein 2013) compared to the material applied for the predecessor part. Figure 2.4 shows examples of these material indices for different materials regarding tension stiffness, sheet bending/buckling and strength in comparison to steel. The higher the relative value of the index, the higher the lightweight

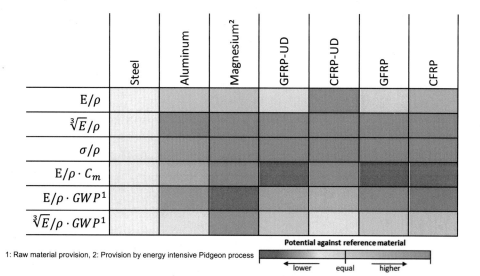

Fig. 2.4 Comparison of material indices considering mechanical performance and density referred to steel

potential of the material. In contrast to the part selection of part candidates (Sect. 1.3), this identification does not only consider the body part to be redeveloped. The focus is especially on alternative materials, geometries or general design approaches.

By a direct comparison of the relevant material indices, an estimation of the resulting material's weight can be given. As one can see, the lightweight potential highly depends on the specific load case. In case of tensile stiffness (E/ρ), for example, only the unidirectional CFRP offers a lightweight potential compared to steel. On the contrary, in case of bending stiffness for sheets or buckling resistance $(\sqrt[3]{E}/\rho)$, each listed material offers a lightweight potential compared to steel.

However, for a bending stiffness problem of a hollow profile (wall thickness small against height and width), the material index to be considers is similar to E/ρ so that the necessary amount of different materials can be estimated in comparison to steel. As can be seen, CFRP, as an example, offers a high lightweight potential against steel. However, this does not give an idea of resulting properties, such as costs (C_m in €) or GWP from raw material provision ($kg\ CO_2\text{-}eq$). By dividing the material index for a specific load case with specific material properties, the corresponding properties of the body part can be estimated.

Taking into account costs and environmental impact, the result shows that most of the lightweight materials have higher costs and a higher environmental impact of the material itself. However, it has to be pointed out that the resulting GWP of the material only represents the raw material production. In order to estimate the aggregate GWP of the entire life cycle (including the manufacturing, the use phase and the end of life), a more comprehensive analysis is necessary including models considering these phases. As the given results still offer a high uncertainty, the additional models are not considered within this level, but within the next level. This short example only represents single materials. However, by this analysis also material combinations, such as multi-layer materials, can be considered.

Out of the selected materials and geometries, there are still many opportunities for general design approaches that lead to different manufacturing routes. The question is, which design approach can be applied regarding the boundary conditions of the usage phase, the bonding to other parts and especially the body part's manufacturing in general. Therefore, possible design approaches are determined out of the selected materials and geometries. For example, a combination of metal, FRP and plastic can, for example, lead to:

- Plastic mould locally reinforced by FRP and/or metal inserts
- FRP with additional plastic mould and metal inserts
- Metal body with plastic moulded ribs and FRP patches

As different design approaches and materials require different boundary conditions, it can be feasible to prove the opportunity for an alteration of the boundary conditions determined on level 0.

2.3.3   Level 2: Generic Topology Model

Level 2 aims at describing the body part to be developed as a generic concept by an analytical determination of properties in two dimensions. By this approach, concept-specific reinforcements and the concepts topology can be taken into account. This leads to an improved estimation of the concept's properties.

The basic principle of this level is an abstraction of the predecessor part. This is done by transferring the part's dimensions and material to a generic model. The actual conceptualization is done on the abstracted level. This allows a direct comparison of the abstracted predecessor and the abstracted concepts with the assumption that the comparison on abstracted and detailed level is similar (see Fig. 2.5). This abstracted model does not take into account specific details of the body part, such as drafts or radii. However, the general geometry and especially the combination of different elements, such as reinforcements, and the specific topology of the concepts can be analysed. In addition, the opportunities of an extended installation space can be analysed in a very early concept phase. The advantages of an analytical calculation are the existing analytical models (e.g. for materials or load cases), a very fast calculation, an easy conceptualization and adaption as well as the opportunity for an automated matching of specific properties.

As no suitable tool for the project's needs was available, a new tool was developed within the project. The focus was on the consideration of the concepts' specific material distribution and topology. For the mechanical properties, the concepts' stiffness, strength and buckling resistance are of major importance. In combination with other properties, such as costs, environmental behaviour or weight, the tool offers a multi-criteria decision-making in order to identify the most promising concepts. This identification seems to be useful for basic variants. The more detailed or complex the concepts get, the more difficult is an appropriate representation by this approach. However, there is already a large number of information that can be gained from the simplified concepts. (Fröhlich et al. 2017).

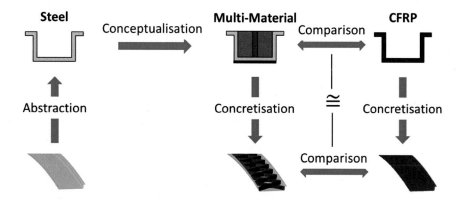

Fig. 2.5   Principle of the conceptualization und comparison of the concepts on abstracted level

In order to describe the predecessor and the concepts to be developed, individual geometrical elements can be combined and predefined or user individual materials can be allocated to them to create multi-material concepts. This includes the allocation of, e.g., multi-layer materials. An example for the stepwise conceptualization is shown in Fig. 2.6 for a u-shape profile by adding additional elements as well as defining the applied materials and their ratio.

On basis of the subsequent calculation of properties, the concepts can be matched to the predecessor's properties, respectively defined requirements. A simplified example for the calculation of the concepts' environmental properties is given in Chap. 6. For the matching of the concept's properties and the development goals, there are two general ways. The first is an automated matching by, for example, adapting the wall thickness of a profile. This way is most suitable for monolithic body parts, as their wall thickness mostly is the only adaptable characteristic. In case of complex concepts, e.g. including anisotropic multi-layer materials, there are much more adjustable variables, which leads to a huge amount of different opportunities. Therefore, an optimization can be used in order to identify the most promising solutions. However, due to different aspects for decision-making that cannot be quantified or directly be represented by the abstracted model, there is still the need for an operation of the designer. The resulting concepts can be compared among each other or to a predecessor either directly by the calculated properties or by a final manual evaluation. By the assessment, the user has the opportunity to consider the prior described not calculable properties and bring in his personal experience.

Depending of the concept specification, different models and tools can be coupled with this tool in order to consider different aspects of the part's life cycle. Examples are:

- Consideration of different general opportunities for manufacturing and their specific material utilization rates (compare Chap. 4)
- Identification of time, energy and resource consumption of specific processes by using general data of production process simulations (compare Chap. 5)
- Different models for the environmental impact of the concept (compare Chap. 6)

Summarizing, this level leads to a first estimation of the concepts mechanical properties in case of stiffness and strength—as well as buckling behaviour for simpler concepts. By

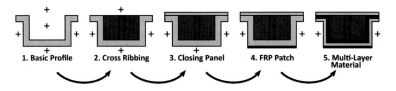

Fig. 2.6 Conceptualization of a profile by addition of geometrical elements and application of individual materials

matching the mechanical properties to the reference, the concepts' weight can be estimated. Basing on the individual materials weights, again, the resulting costs and environmental impact can be estimated including a general consideration of manufacturing. Despite the remaining uncertainty by this abstracted definition, the results allow a first estimation of the concepts' properties with low effort. In addition, the accuracy can be adapted by using different data and models for estimating the concepts' properties.

2.3.4 Levels 3 and 4: Detailed Concept Model and System Model

On level 3, the previously developed concepts are designed more detailed in CAD and analysed by a FE-Simulation as it is common practice in industry. These detailed models allow a consideration of more complex geometries, local characteristics (e.g. beadings), more specific load applications with local effects and especially the effect of more detailed manufacturing constraints. This also allows more comprehensive mechanical analyses including local stresses/strains, which cannot be sufficiently be described on level 2, a more detailed manufacturing process planning and simulation as well as more specific application models for the calculation of the environmental properties.

As the effort for concept creation and analysis as well as for changes is high in comparison to the abstracted analytical models, only a limited amount of concepts can be assessed compared to level 2. However, the detailed design leads to a better identification of the individual process steps and parameters. This again leads to a more specific consideration of costs and environmental impact.

As the concept to be developed is not independent, but only a part of a superordinate system, level 4 considers the entire system behaviour. This concerns, for example, an increase of loads on the considered body part, due to an increased stiffness. In addition, the behaviour of the body part in case of a crash depends on its integration into the entire system and vice versa; the behaviour of the entire vehicle depends on the individual part and their assembly.

Compared to level 3, the analysis of the entire system especially considering crash load cases leads to a much higher effort. This is caused by, for example, the necessary installation of the body part into the entire system model and the intensive simulation including pre- and postprocessing. Therefore, only concepts should be analysed in this step, which have the chance to be feasible after the analyses on the prior steps.

2.4 Case Study

In this case study, the before-presented procedure is exemplary applied to two body parts considered within the project. Within the case study, the number of evaluation criteria is reduced to the resulting concepts' weight, costs and GWP with mechanical performance matched to the demanded development goals.

Both body parts are out of the SLC, but show a very different applicability of the early abstracted level of the procedure, because of their very different mechanical boundary conditions. The materials applied in the two reference parts lead to different scenarios for the development of concepts. The considered body parts within the project and within the case study in particular are shown in Fig. 2.7.

In order to be able to compare the predecessor and the developed concepts, the properties of the predecessor have not been taken over from the SLC, but have been calculated basing on consistent data used in the project.

2.4.1   Case Study 1: Roof Reinforcement

The first considered body part is the roof reinforcement, which is placed between the two b-pillars of the SLC, as it is applied in most of the state-of-the-art vehicles. The main task of the roof reinforcement is, on the one hand, to support the two upper ends of the b-pillars to increase the vehicle body's stiffness and in case of a side crash. On the other hand, it is connected to the roof by a non-structural bond in order to increase the roof sheet's buckling resistance.

The roof reinforcement is made out of a 0/90° epoxy-CFRP manufactured by pultrusion. This results in a low weight, but high costs. As CFRP has a very high environmental impact in case of CO_2 emission, the resulting Global Warming Potential (GWP) of the manufacturing (including raw material production) is very high.

After the analysation of the predecessor, the requirements for the concepts to be developed are defined. Besides a reduction of costs, the most important goal by developing new concepts for the roof reinforcement is to reduce the GWP compared to the predecessor part. Therefore, a certain increase of weight is accepted, as long as the resulting increase of GWP in the use phase can be compensated by the decrease in raw material production and manufacturing.

As the bending stiffness and stability have the greatest influence on the expected mechanical behaviour, a decrease of the concepts' torsional stiffness is accepted as well. This allows an application of open profiles, such as a u-shape profile, which can hardly reach the torsional stiffness of a hollow profile, such as given for the predecessor part.

Centre Tunnel   Roof Reinforcement

Fig. 2.7 Parts considered in the case study (extracted from Volkswagen, 2019)

Fig. 2.8 Properties of the considered materials (raw material production only) compared to the epoxy-CFRP applied in the predecessor part.

In case of the density-depended mechanical properties, no material, except the alternative CFRP, shows an advantage compared to the reference material. However, taking into account the costs and CO_2-emission, the analysis reveals a huge potential for reducing both by accepting an increasing weight. Therefore, no of these general materials can be excluded.

In this example, only properties of one specific material are shown to give an overview. Depending on heat treatment, alloying or fibre ratio, each material offers a wide range of properties. In order to compensate the specific weak points of the single materials compared to the reference material, more specific materials and alternative process routes have to be analysed. Therefore, a huge amount of material selection tools can be used as already mentioned in Sect. 2.2.

The developed basic geometry variants consider a general design approach, and the stepwise development of different variants for each basic geometry is exemplary shown in Fig. 2.9. All general concepts have been further detailed by application of different materials as well as different reinforcement strategies, such as injection moulded ribs, FRP patches or multi-layer materials.

The analyses on levels 2 and 3 revealed very similar results, as the roof reinforcement's geometry and load cases can sufficiently be described by an analytical 2D model. In case of its mechanical performance and especially its adaptability, the u-shape profile has been identified as the most promising solution. However, this basic concept offers a

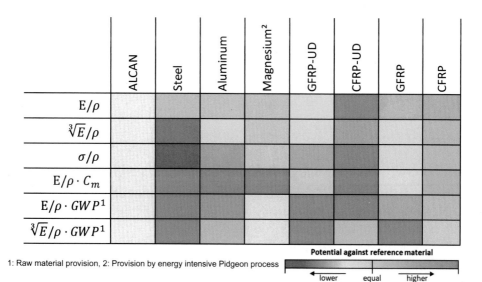

Fig. 2.8 Comparison of different materials regarding the reference material of the roof reinforcement for selected material indices (raw material only)

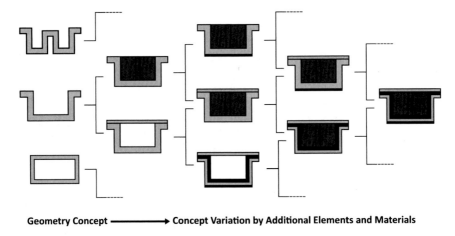

Geometry Concept ——————▶ **Concept Variation by Additional Elements and Materials**

Fig. 2.9 Example for the development of general design concepts by variation of a basic geometry concept

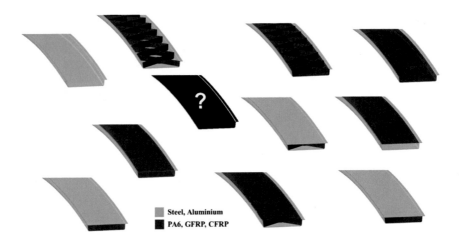

Fig. 2.10 Extract of possible design variants for a u-shape roof reinforcement with and without closing panel

multitude of different specific designs including different material applications and an additional closing panel as shown exemplary in Fig. 2.10. Therefore, it can be adapted to different scenarios by creating adapted variants.

The application of a closing panel improves the mechanical performance of the roof reinforcement in case of torsion as well as bending. However, the additional process step increases the resulting costs. Furthermore, the wall thickness has to be decreased for the same mechanical performance. In this scenario, this leads to a thin wall thickness for the application of steel. This results in a decreased resistance against buckling respectively collapsing. Therefore, for this scenario steel is only applied on u-shape profiles without closing panel. This is an example of the scenario dependent applicability of different materials or design approaches.

The analysation on level 4 is done by analysing the vehicle's bending and torsional stiffness as well as a side crash. As expected, the steel concepts with a closing panel failed the crash simulation due to a collapse close to the connection to the body in white. Other concepts with a higher wall thickness resist the crash without failure.

In this example, most of the concepts' specification could already be done on more abstracted models. On the one hand, the body part's geometry is quite simple. On the other hand, the relevant load cases can sufficiently be described by surrogate models.

2.4.2 Case Study 2: Centre Tunnel

The second body part is a centre tunnel, which is made out of a tailored blank hot forming steel with a higher wall thickness on the upper part. The result is a medium to high weight and medium manufacturing costs. As steel has a low environmental impact in case of CO_2 emission, the resulting Global Warming Potential (GWP) of the manufacturing (including raw material production) is very low. As the reference has a high potential of weight reduction, the development goal is to reduce the weight by 20%. However, as a change of material from steel to lightweight materials usually comes along with an increase of costs and environmental impact for the production, the goal is not to exceed 110% of the reference costs and environmental impact.

The centre tunnel is placed in the middle of the vehicle in order to allow a transmission of the exhaust system. Furthermore, the centre tunnel protects the passengers from the environment especially in case of crash load cases. Particularly relevant are front and side crashes. So the tunnel's load conditions differ a lot from the roof reinforcement, which can be treated as a bending beam on the first levels. As the crash behaviour of a body part can only be verified in a crash simulation including the entire vehicle, the early level of concept development plays a subordinate role. The feasibility of different materials as well as a necessary wall thickness also depends on local stresses and strains. However, a general feasibility of certain materials or design approaches in case of resulting costs or environmental impact can be given nonetheless.

In order to get a first impression of the performance of general materials related to the reference material, Fig. 2.11 shows the results for selected material indices of related to the reference material.

The result shows that developing multi-material concepts considering the developing goals offers significant challenges especially regarding the concepts costs and environmental impact. Especially an intensive application of CFRP might exceed the given limits. The results for the analytical estimation basing on generic geometries and generic load cases show similar results. Depending on the specific material, aluminium only shows slight opportunities for a weight reduction. On the contrast, CFRP offers a high

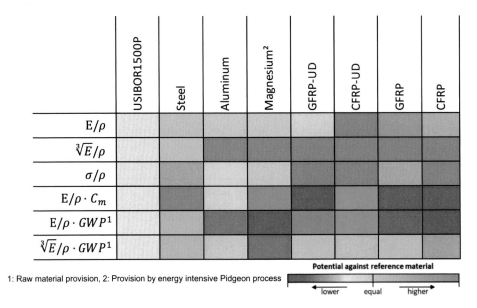

Fig. 2.11 Comparison of different materials regarding the reference material of the centre tunnel for selected material indices (raw material only)

potential for weight reduction, but only by significantly higher costs and environmental impact. However, these considerations do not consider high strain rates and the necessary elongation behaviour in case of a crash.

Basing on the preliminary investigations, a large number of different concepts including multiple variants have been developed. All concepts are matched to the expected properties regarding stiffness as well as crash load cases. The main evaluation criterion has been the intrusion of the bulkhead into the passenger compartment caused by a front crash without a failure of the body part.

Fig. 2.12 shows an extract of the developed concepts and the resulting properties in the examples of weight, life cycle environmental impact and costs related to the given development goal. As the shown tunnels are still on a concept level, there is an uncertainty in the determined property values. The calculation of costs bases on a green-field scenario including the investment for the necessary production facilities as an additional value for the comparison beside the manufacturing costs. This ensures a comparability of different concepts with different manufacturing routes. The manufacturing itself was calculated by considering the raw material, a detailed analysis of each manufacturing step as well as the specific resulting overhead costs for each concept. The environmental impact in this example bases on the green house potential (CO_2-equ.) as key indicator for the environmental impact. Therefore, the raw material provision and manufacturing of the tunnel is considered as well as a usage scenario of a class A vehicle (e.g. VW Golf VII) with gasoline turbocharged drivetrain and a mileage of 200,000 km (WLTC Cycle).

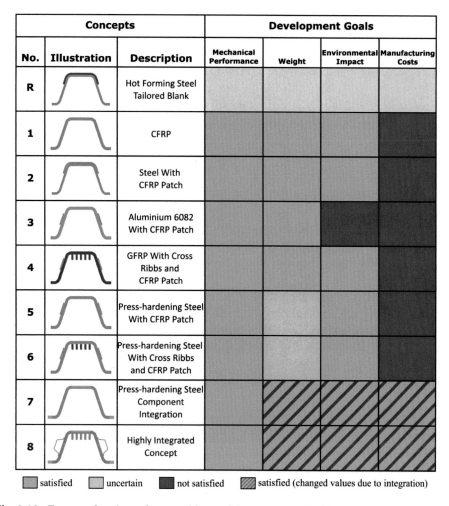

Concepts			Development Goals			
No.	Illustration	Description	Mechanical Performance	Weight	Environmental Impact	Manufacturing Costs
R		Hot Forming Steel Tailored Blank				
1		CFRP				
2		Steel With CFRP Patch				
3		Aluminium 6082 With CFRP Patch				
4		GFRP With Cross Ribbs and CFRP Patch				
5		Press-hardening Steel With CFRP Patch				
6		Press-hardening Steel With Cross Ribbs and CFRP Patch				
7		Press-hardening Steel Component Integration				
8		Highly Integrated Concept				

satisfied   uncertain   not satisfied   satisfied (changed values due to integration)

Fig. 2.12 Extract of variants for a multi-material centre tunnel with their mechanical performance, weight, environmental impact and costs compared to the reference part

A more detailed presentation of the LCA is shown in Chap. 6. In order to investigate the deviations of the results of the different level, also unfavourable concepts have been further investigated. The colour code for the fulfilment of the development goals distinguishes between as follows:

- Green: Goal is satisfied
- Yellow: Satisfaction is uncertain as the property's value of the concept is close to the goal's minimum or maximum
- Red: Goal is not satisfied
- Shaded: Different value of the development goal due to integration of parts

The concepts 1 to 6 follow a sole material substitution strategy with maintained system boundary of the reference tunnel. As concepts including only a single material generally offer weak points regarding specific criteria, the focus was on the application of multiple materials. However, concept 1 is an example for a single material concept out of CFRP organosheet. The application of CRFP results in a very good mechanical performance at very low weight. The significant weight reduction is why the environmental impact of concept 1 meets the goal, despite CFRP's high environmental impact. However, the high costs especially of carbon fibres lead to unsatisfied manufacturing costs. The concepts 2 to 6 consider the application of a material in combination with local reinforcements of CFRP patches. Except concept 3, all concepts meet the environmental impact goals. However, due to the more complex manufacturing as well as higher material costs of the applied lightweight materials, the manufacturing costs do not match the given goal. Finally, no material substitution concept is able to meet all given strict development goals. Therefore, in addition to the material substitution of the prior concepts, the concepts 7 and 8 integrated additional adjacent body parts. This aimed at increasing the design freedom of the concepts as well as decreasing the manufacturing and especially assembly effort. Due to the integration of additional parts to the system boundary, the development goals are enhanced by the integrated component's properties. As can be seen, both integrated concepts are able to meet the development goals, as the integration allow a higher design freedom and a significant reduction of manufacturing effort.

The results of the concepts' analyses show that most of the concepts missed the given development goals of this scenario especially in case of manufacturing costs. This is a consequence of higher material costs as well as additional manufacturing steps. However, applied to another scenario with, for example, a more FRP-intensive reference, these concepts can be a competitive and more environmental friendly alternative. In contrast, the concepts considering a part integration comply with all development goals. This shows that an applicability of multi-material design especially including an application of FRP highly depends on the given scenario. In addition, a pure material substitution does not exploit the full potential of multi-material design due to the mostly higher costs and higher environmental impact of lightweight materials in comparison to steel.

As expected, the results of level 2 and level 4 show a significant difference regarding the exact values of the resulting properties. Especially local properties could not sufficiently been considered in the early steps, which nevertheless have a huge influence on the feasibility of the concepts. However, in case of the general costs and environmental impact, specific challenges could be identified in the early steps.

2.5   Summary and Outlook

This chapter presented an approach for multi-material design body parts and the application of this approach by developing several concepts for a given scenario. Basing on the Extended Mapping Matrix Approach by Köhler, different levels of concept development were defined. Therefore, different product models have been used that take different level of abstraction in the concept development into account. In a case study, a multitude of different concepts for a roof reinforcement and a centre tunnel were presented. It was shown that the applicability of the early and the resulting information highly depends on the specific development task. Especially the complexity of the specific geometry and load cases offer a disparate representability of the different models. In addition, the feasibility and especially competitiveness of multi-material design depends on the given development scenario regarding the predecessor part and resulting development goals. As shown by the concepts of the centre tunnel, a part integration strategy offers high a potential for a reduction of weight, costs and environmental impact. This is caused by a higher degree of freedom in design as well as lower manufacturing costs as less individual process steps are necessary.

In order to improve the applicability of the presented development approach for varying development scenarios, additional product models have to be integrated. As the presented levels are only of exemplary nature, there is still an amount of opportunities for other product models in between the presented level. This leads to a better predictability concerning different body parts and boundary conditions. In case of multi-material design, there is a need to exploit the opportunities of the creation of variants. Especially local reinforcements can be applied or adapted to the specific use case and enable a low-effort creation of variants. In order to exploit the full potentials of multi-material design, it is necessary to consider all the additional properties of the applied materials. Therefore, an integration of additional non-structural functions into structural body parts can be a major step for an economical application of multi-material design as shown by (Klaiber et al. 2019) or (Fröhlich et al. 2019).

References

Ashby, M. F.; Bréchet, Y.J.M.; Cebon, D.; Salvo, L. (2004): Selection strategies for materials and processes. In: Materials & Design 25 (1), S. 51–67. DOI: https://doi.org/10.1016/s0261-3069(03)00159-6.

Ashby, Michael F. (2005): Materials Selection in Mechanical Design. 3. Aufl. s.l.: Elsevier professional. Online verfügbar unter http://gbv.eblib.com/patron/FullRecord.aspx?p=288921.

Athawale, Vijay Manikrao; Kumar, Rajanikar; Chakraborty, Shankar (2011): Decision making for material selection using the UTA method. In: Int J Adv Manuf Technol 57 (1–4), S. 11–22. DOI: https://doi.org/10.1007/s00170-011-3293-7.

Babanli, M. B.; Prima, F.; Vermaut, P.; Demchenko, L. D.; Titenko, A. N.; Huseynov, S. S. et al. (2019): Material Selection Methods: A Review. In: Rafik A. Aliev, Janusz Kacprzyk, Witold Pedrycz, Mo Jamshidi und Fahreddin M. Sadikoglu (Hg.): 13th International Conference on Theory and Application of Fuzzy Systems and Soft Computing—ICAFS-2018, Bd. 896. Cham: Springer International Publishing (Advances in Intelligent Systems and Computing), S. 929–936.

Bader, Benjamin; Türck, Eiko; Vietor, Thomas (2019): MULTI MATERIAL DESIGN. A CURRENT OVERVIEW OF THE USED POTENTIAL IN AUTOMOTIVE INDUSTRIES. In: Klaus Dröder und Thomas Vietor (Hg.): Technologies for economical and functional lightweight design. Conference proceedings 2018. Berlin: Springer Vieweg (Zukunftstechnologien für den multifunktionalen Leichtbau), S. 3–13.

Dröder, Klaus (Hg.) (2020): Prozesstechnologie zur Herstellung von FVK-Metall-Hybriden. Ergebnisse aus dem BMBF-Verbundprojekt ProVorPlus. 1. Auflage 2020. Berlin: Springer Berlin; Springer Vieweg (Zukunftstechnologien für den multifunktionalen Leichtbau).

Duflou, J. R.; Moor, J. de; Verpoest, I.; Dewulf, W. (2009): Environmental impact analysis of composite use in car manufacturing. In: CIRP Annals 58 (1), S. 9–12. DOI: https://doi.org/10.1016/j.cirp.2009.03.077.

Fröhlich, Tim; Klaiber, Dominik; Türck, Eiko; Vietor, Thomas (2019): Function in a box: An approach for multi-functional design by function integration and separation. In: Procedia CIRP 84, S. 611–617. DOI: https://doi.org/10.1016/j.procir.2019.04.343.

Fröhlich, Tim; Kleemann, Sebastian; Türck, Eiko; Vietor, Thomas (2017): Multi-criteria analysis of multi-material lightweight components on a conceptual level of detail. DOI: https://doi.org/10.24355/DBBS.084-201709070942.

Giaccobi, S.; Kromm, F. X.; Wargnier, H.; Danis, M. (2010): Filtration in materials selection and multi-materials design. In: Materials & Design 31 (4), S. 1842–1847. DOI: https://doi.org/10.1016/j.matdes.2009.11.005.

Goede, Martin; Stehlin, Marc; Rafflenbeul, Lukas; Kopp, Gundolf; Beeh, Elmar (2009): Super Light Car—lightweight construction thanks to a multi-material design and function integration. In: Eur. Transp. Res. Rev. 1 (1), S. 5–10. DOI: https://doi.org/10.1007/s12544-008-0001-2.

Götz, Peter; Reg, Yvonne; Hühn, Dominic; Roth, Stephan; Masseria, Frederic; Bublitz, Dennis (2018): Qualitätsgesicherte Prozesskettenverknüpfung zur Herstellung höchstbelastbarer intrinsischer Metall-FKV-Verbunde in 3D-Hybrid-Bauweise – Q-Pro. Unter Mitarbeit von TIB-Technische Informationsbibliothek Universitätsbibliothek Hannover und Technische Informationsbibliothek (TIB).

Jahan, A.; Ismail, M. Y.; Sapuan, S. M.; Mustapha, F. (2010): Material screening and choosing methods – A review. In: Materials & Design 31 (2), S. 696–705. DOI: https://doi.org/10.1016/j.matdes.2009.08.013.

Kaluza, Alexander; Kleemann, Sebastian; Broch, Florian; Herrmann, Christoph; Vietor, Thomas (2016): Analyzing Decision-making in Automotive Design towards Life Cycle Engineering for Hybrid Lightweight Components. In: Procedia CIRP 50, S. 825–830. DOI: https://doi.org/10.1016/j.procir.2016.05.029.

Kellner, Philipp (2014): Zur systematischen Bewertung integrativer Leichtbau-Strukturkonzepte für biegebelastete Crashträger. 1st ed. Göttingen: Cuvillier Verlag. Online verfügbar unter https://ebookcentral.proquest.com/lib/gbv/detail.action?docID=5018942.

Klaiber, Dominik; Fröhlich, Tim; Vietor, Thomas (2019): Strategies for function integration in engineering design: from differential design to function adoption. In: Procedia CIRP 84, S. 599–604. DOI: https://doi.org/10.1016/j.procir.2019.04.344.

Kleemann, Sebastian; Fröhlich, Tim; Türck, Eiko; Vietor, Thomas (2017): A Methodological Approach Towards Multi-material Design of Automotive Components. In: Procedia CIRP 60, S. 68–73. DOI: https://doi.org/10.1016/j.procir.2017.01.010.

Klein, Bernd (2013): Leichtbau-Konstruktion. Wiesbaden: Springer Fachmedien Wiesbaden.

Köhler, Christian; Conrad, Jan; Wanke, Sören; Weber, Christian (2008): A matrix representation of the CPM/PDD approach as a means for change impact analysis. In: Design 2008: proceedings of the 10th International Design Conference, Dubrovnik, Croatia, May 19–22, 2008 / Ed.: Dorian Marjanovic. – Zabreb: Faculty of Mechanical Engineering and Naval Architecture, University of Zagreb, 2008. DOI: https://doi.org/10.22028/D291-22502.

Kuhn, Christoph; Klaiber, Dominik; Altach, Johannes (2019): Lightweight Design in the Vehicle Structure with the Example of a Center Tunnel for the Porsche Boxster. In: Kunststoffe international 2019 (04). Online verfügbar unter https://www.kunststoffe.de/en/journal/archive/article/lightweight-design-in-the-vehicle-structure-with-the-example-of-a-center-tunnel-for-the-porsche-boxster-7979083.html.

Sakundarini, Novita; Taha, Zahari; Abdul-Rashid, Salwa Hanim; Ghazila, Raja Ariffin Raja (2013): Optimal multi-material selection for lightweight design of automotive body assembly incorporating recyclability. In: Materials & Design 50, S. 846–857. DOI: https://doi.org/10.1016/j.matdes.2013.03.085

Song, Young S.; Youn, Jae R.; Gutowski, Timothy G. (2009): Life cycle energy analysis of fiber-reinforced composites. In: Composites Part A: Applied Science and Manufacturing 40 (8), S. 1257–1265. DOI: https://doi.org/10.1016/j.compositesa.2009.05.020.

Wanner, Alexander (2010): Minimum-weight materials selection for limited available space. In: Materials & Design 31 (6), S. 2834–2839. DOI: https://doi.org/10.1016/j.matdes.2009.12.052.

Weber, Chr (2007): Looking at "DFX" and "Product Maturity" from the Perspective of a New Approach to Modelling Product and Product Development Processes. In: Frank-Lothar Krause (Hg.): The Future of Product Development. Berlin, Heidelberg: Springer Berlin Heidelberg, S. 85–104.

Weber, Julian (2009): Automotive Development Processes. Berlin, Heidelberg: Springer Berlin Heidelberg.

Knowledge Management

3

Sebastian Kleemann

Abstract

Multi-material-design is a promising approach for the automotive industry to operate economical lightweight design. Combining metallic materials with composites and short-fibre reinforced plastics enables to develop lightweight components with superior mechanical properties. One of the central challenges is that most car body developers in the automotive industry have gathered little experience with plastics, composites and multi-material designs. One approach of product development to meet this challenge is the provision of knowledge, for example through design rules and design principles. The scope of this chapter is to provide knowledge for developing multi-material designs. First, relevant knowledge is identified, and an access concept for the design rules is developed. Furthermore, what will be developed is a text similarity algorithm for identifying similar rules. Afterwards, the knowledge management system will be prototypically implemented and applied to two case studies.

3.1 Introduction

Vehicle developers, especially car body developers, have a wealth of experience in developing conventional steel or aluminium car bodies. However, they have limited experience with composite design and multi-material designs. Therefore, one might conclude

S. Kleemann (✉)
Institute for Engineering Design (IK), Technische Universität Braunschweig, Braunschweig, Deutschland
e-mail: s.kleemann@tu-braunschweig.de

© Springer-Verlag GmbH Germany, part of Springer Nature 2023
T. Vietor (ed.), *Life Cycle Design & Engineering of Lightweight Multi-Material Automotive Body Parts*, Zukunftstechnologien für den multifunktionalen Leichtbau,
https://doi.org/10.1007/978-3-662-65273-2_3

that the current lack of knowledge for multi-material design is one major challenge. This leads to the conclusion that the designer must be provided with extensive knowledge in order to be able to successfully develop multi-material designs. This required knowledge includes, among other things, the material behaviour of the various materials, their production technology and the applicable joining technology. However, this knowledge is already available in other industries such as aerospace and has at least been published in excerpts. This applies above all to the specific knowledge of composites. These materials are currently used primarily in the aerospace industry, and what is incompatible with the requirements of the automotive industry in terms of process speed and costs is the complexity of design, manufacturing and quality assurance methods. This knowledge is therefore not readily transferable to automotive engineering. An approach to address a missing knowledge base can be found in methodical product development in the form of design rules (Inkermann et al. 2017; Roth 2000; Ziebart 2012).

Extensive literature is devoted to the subject lightweight design. Within the scope of this work, lightweight design is understood as a declaration of intent to reduce weight for functional or economic reasons without reducing the function of a product. Based on this, different degrees of lightweight design can be defined, which describe the consequence with which the above-mentioned intention to reduce weight is implemented.

The product developer can meet the actual lightweight design with different lightweight design strategies, which attack on different abstraction levels (requirements, functions, principles, shape). Based on the understanding that design is a change from analysis and synthesis and that the results of synthesis can only be as good as the underlying knowledge base, there is a need to provide knowledge. (Hatchuel 2003; Weber 2007).

3.2 Review on Knowledge Management in Product Development

Product development is always an interplay between analysis and synthesis (Hatchuel 2003; Weber 2007), and that synthesis is decisively influenced by the available knowledge base. This is why, this chapter discusses how knowledge can be provided. First, the fundamentals of knowledge management are explained, with particular emphasis on the knowledge management process. Subsequently, specific approaches to providing knowledge are explained.

Knowledge management encompasses all strategic and operational activities and management tasks that aim at an ideal handling of knowledge. Knowledge can be distinguished in tacit knowledge and explicit knowledge (Snowden 2002). Tacit knowledge represents internalized knowledge that an individual may not be consciously aware of, such as to accomplish particular tasks. At the opposite end of the spectrum, explicit knowledge represents knowledge that the individual holds consciously in mental focus, in a form that can easily be communicated to others. Explicit knowledge can be

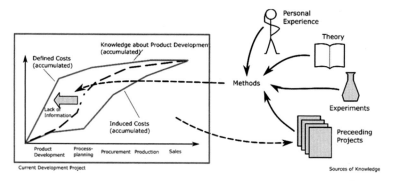

Fig. 3.1 Sources, types and access to knowledge during product development, based on (Nowak 1997; Vajna 2001) from (Inkermann et al. 2017)

described and consequently stored in a machine-readable form. Implicit knowledge cannot be codified or brought into a codifiable form with justifiable effort.

Knowledge plays a special role in product development. On the one hand, only those concepts can be developed for which the necessary knowledge base is available (see above), on the other hand knowledge also enables directional decisions. Decisions, especially at the beginning of product development, have a great influence on costs and are often made based on incomplete information. However, wrong decisions may only be uncovered in the course of development and may only be corrected by considerable additional costs (Ehrlenspiel 2014; Roth 2000). In order to counter this situation, companies try to support their processes through frontloading. The required information is based on various sources and must be timely, reliable, comprehensible, and adapted, prepared and made available to the specific boundary conditions of the respective development task (Nowak 1997). (see Fig. 3.1).

3.2.1   Approaches to Knowledge Provision

Design rules are one of the main sources of explicit knowledge from engineering practice (Dieter 1991). The main sources for design rules are literature, direct experience of practising developers and established design practices of companies. Edwards defines the term design rules as follows: "A design rule is a principle that is proposed to set standards or establish a procedure" (Edwards 1993). Design rules are explicit knowledge about engineering practice. They are often based on successful designs and are therefore increasingly specific to a particular domain and can cover a wide range of experience in the use of existing technology. A specific implementation of design rules in product development was developed within the framework of the research project HI-PAT\ at the Institut für Konstruktionstechnik of Technische Universität Braunschweig (Nehuis 2011). Fig. 3.2 shows the knowledge representation in spreadsheet format.

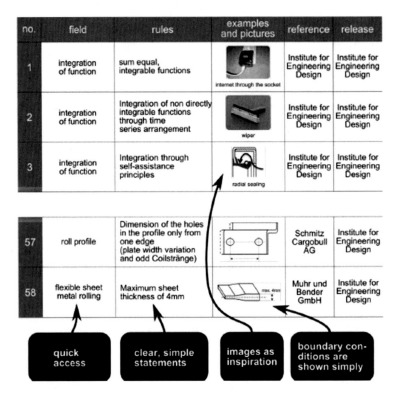

no.	field	rules	examples and pictures	reference	release
1	integration of function	sum equal, integrable functions	*internet through the socket*	Institute for Engineering Design	Institute for Engineering Design
2	integration of function	Integration of non directly integrable functions through time series arrangement	*wiper*	Institute for Engineering Design	Institute for Engineering Design
3	integration of function	Integration through self-assistance principles	*radial sealing*	Institute for Engineering Design	Institute for Engineering Design
57	roll profile	Dimension of the holes in the profile only from one edge (plate width variation and odd Coilstränge)		Schmitz Cargobull AG	Institute for Engineering Design
58	flexible sheet metal rolling	Maximum sheet thickness of 4mm		Muhr und Bender GmbH	Institute for Engineering Design

quick access | clear, simple statements | images as inspiration | boundary conditions are shown simply

Fig. 3.2 Detail of the structured design rules with a clear statement, the corresponding image and the references according to (Nehuis et al. 2011)

This implementation offers a systematic structure and search and filter functions for design rules. The collection of design rules can be filtered with the following key topics:

- Function integration,
- Lightweight design,
- Roll forming,
- Tailored Rolled Blanks,
- Material.

Each design rule is defined with a clear statement and a corresponding image and references (see Fig. 3.2). The instructions allow quick access to the knowledge. The images of the design rules allow even inexperienced product developers to quickly grasp complex situations (Salustri et al. 2008; Mayer et al. 1990), where, for example, van Wie shows that a negative and a positive example improves the understanding of the rule (van Wie 2002; Nehuis 2011).

Design rules must be practical and easy to use. A design rule is effective if it can be learned and applied with little effort and if it improves the quality and quantity of the results. However, this effectiveness is strongly influenced by the framework conditions, such as the complexity of the task and the time and resources required. In addition, a rule must be technically complete and comprehensive. Ideally, a design rule reduces the number of iterations and thus development time and costs. Furthermore, many product developers are familiar with handling design rules. This increases acceptance, shortens training time, increases the ability to remember and facilitates application. In addition, design rules can be created inexpensively in an appropriate form and are easy to distribute.

Bischof has compiled the following requirements for design rules (Bischof 2010; Mayer 1990; van Wie 2002):

- mainly graphical with supporting texts.
- using short texts that are as clear and understandable as possible.
- matching the supporting function of the illustration with the complexity of the text and its information.
- using (self-)descriptive illustrations and naming all components of the system and their relationships.
- consistent and concise formulation of the knowledge.
- formulation independent of the product development methodology.
- formulate as explicit a design note as possible to show how the rules are applied.
- justifying the design rule
- matching the illustration and texts to the background knowledge of the users.

3.3   Providing Knowledge for Multi-Material Designs

Developing multi-material lightweight automotive body parts requires extensive specialist knowledge on various topics. A knowledge management system is being developed to accelerate the development of components in multi-material design and to expand the know-how of the engineers. Multi-material design or at least related disciplines are the subject of numerous publications. This knowledge is often tailored to specific applications and industries and cannot easily be transferred to vehicle development. However, the knowledge management system should include general rules and recommendations for lightweight design and specific design rules for production processes and materials. Fig. 3.3 shows the conceptual structure of the design rules for multi-material design.

As explained in the state of the art, the design rules consist of comparative illustrations of an advantageous and disadvantageous design as well as explanatory text and a title. Furthermore, the concept envisages different structuring options as well as sources and similar rules.

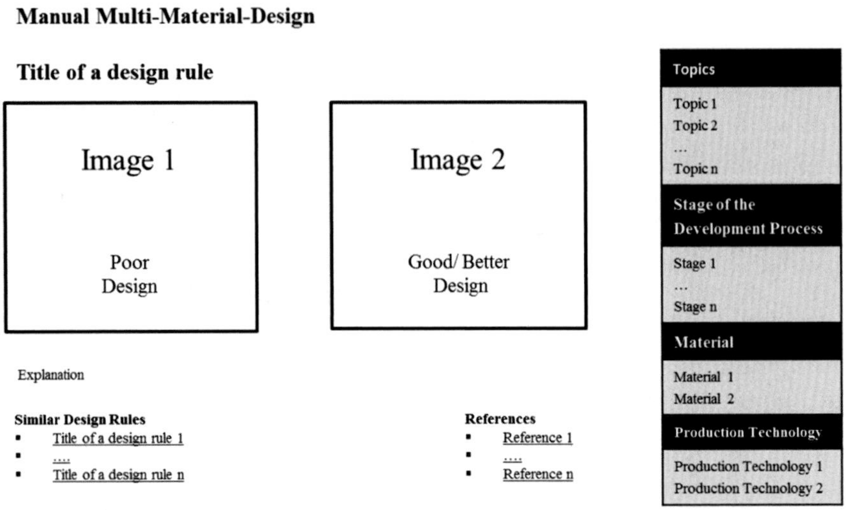

Manual Multi-Material-Design

Title of a design rule

		Topics
Image 1	Image 2	Topic 1 / Topic 2 / ... / Topic n
		Stage of the Development Process
Poor Design	Good/ Better Design	Stage 1 / ... / Stage n
		Material
Explanation		Material 1 / Material 2
Similar Design Rules • Title of a design rule 1 • .... • Title of a design rule n	**References** • Reference 1 • .... • Reference n	**Production Technology** Production Technology 1 Production Technology 2

Fig. 3.3 Conceptual structure of the design rules for multi-material construction based on Kleemann et al. (2016)

3.4   Identifying Relevant Knowledge

Based on the collection of design rules and principles from the BMBF project HIPAT (cf. State of the Art), the manual multi-material design will be developed. The design rules are extended by new topics such as composites, joining technology, forming and injection moulding. The identification of relevant knowledge is divided into three sections, which explain different sources of knowledge, the literature, current applications of multi-material construction and the derivation of new design rules from own developments.

3.4.1   Identify Relevant Knowledge from the Literature

The starting point for the design rules is technical literature. In the first step, existing design rules and principles are researched and prepared. These include, among others, the design rules for composites according to (Schürmann 2007), the lightweight design principles according to (Klein 2013) as well as rules for injection mouldable designs from (Brinkmann 2011). It was important at this point to prepare all rules and principles in such a way that there are comparative illustrations and short explanatory texts that use consistent vocabulary.

3.4.2 Identify Relevant Knowledge Based on Current Applications of Multi-Material Design

Most development projects rely on existing knowledge. The reuse of established solutions reduces the development risk and costs, as the properties of the solution elements can be known or determined with moderate effort. Based on a literature research of current hybrid design application from automotive engineering, the essential potentials are defined and design principles for multi-material construction are derived.

Table 3.1 shows an excerpt from the analysis. The examined components are mainly structural components such as b-pillars, attachments such as frontends or front end carriers and flaps from research and development projects that have already been completed or are currently underway. The components are either parts that have already been used in existing vehicles, or pre-series or research projects to demonstrate specific technologies without using these parts in industry. Table 3.1 contains three examples of analysed

Table 3.1 Excerpt of the list of analysed components from (Kleemann et al. 2017b)

Component	Frontend	B-Pillar	Roof Crossmember
Readiness Level	Serial part. Audi A8	Pre-Serial part, Research TU Dresden	Serial part: Audi A6
Material Combination	Aluminium sheet metal PA6 GF30 Continuous filament organic sheet (glass fibre, PA6 matrix)	Steel sheet metal Glass LFT with PA matrix Continuous filament organic sheet metal (glass fibre, PA6 matrix)	Steel sheet metal PA6 GF30
Structure	Aluminium body sheet connected to the organic sheet lower belt with PA6 GF30 by injection moulding process	Steel body sheet reinforced with organic sheet and glass LFT with PA matrix by impact extrusion process	Steel body sheet reinforced with PA6 GF30 by injection moulding process
Joining Technology	Form closure In mould lamination	In mould lamination	Form closure In mould lamination
Main Potentials	• Weight reduction (−20%) on lower belt • Reduced assembly	• Weight reduction (−10%) at equal failure behaviour • One shot manufacturing process	• Weight reduction (−30%) at equal component costs • Reduction of components (one part less)
Reference	Lanxess 2010	Gude et al. 2015	Jäschke and Dajek. 2004

components and collects exemplary information. The analysis was structured by two general questions:

- How is the multi-material design implemented in the specific application?
- What are the goals and reasons stated by manufacturers and researchers for using multi-material designs?

The information to answer these questions is compiled from articles in trade journals, press releases and trade fair appearances of the manufacturers or the participating research institutions. The survey evaluated the readiness for series production of the components, the material combination used, the design as well as the joining technologies within the component.

This information characterizes the solution itself and provides insights into the derivation of problem-oriented and procedural knowledge to support multi-material design. The analysis of the solutions itself provides information on how multi-material-design may be implemented. In order to support the realization of components on the basis of multi-material design, design guidelines are formulated with regard to the existing solutions.

3.5   Accessing Relevant Knowledge

To demonstrate the potential of the emerging multi-material design, most parts can be classified as first level multi-material designs, which are a combination of metal and plastic within a component. Various mechanisms give access to design rules and principles. On the one hand, the design rules are assigned to phases of the development process, materials and manufacturing technologies. In this way, the user finds the relevant design rules more quickly.

In addition, the structuring characteristics are linked to each other, for example, materials are assigned to a selection of manufacturing technologies. These manufacturing technologies are also linked to manufacturing machines, since these determine producible geometries and material combinations. In addition, with regard to industrial applications, it is essential to take existing production machines into account, as each production machine has its specific limits with regard to component size and realizable dimensions. This allows the user to find specific design rules for materials or production technologies. The potential model for multi-material designs serves as an additional access concept. The potential model should show not yet exhausted possibilities of hybrid design. These potentials concern, e.g., product characteristics like weight or number of pieces as well as process-dependent characteristics like assembly effort or manufacturing time.

Accordingly, not only design rules and principles were derived from the literature research of implemented hybrid design, but the desired improvements were also documented. Unfortunately, the authors were not being able to check whether these improvements had actually been achieved. This information, however, provides knowledge on a more strategic level which deals with the question why multi-material design should be used.

Based on the analysis of 56 components, main potentials were derived and divided into seven areas. The detailed analysis and derivation of potentials maybe found in Kleemann et al. (2017b). The resulting potential model is shown in Fig. 3.4. The outer ring of the model shows measures for realizing different potentials. More than one measure was derived for each potential. In addition, there are measures that address more than one field of potentials. For example, "Improving local stiffness" is a measure that is clearly assigned to the area "Improving properties and/or functions", while "Integration of joints" is a measure assigned to the areas "Extended design freedom" and "Reduction of interfaces and assembly effort". The table gives an overview of the defined potentials, the assigned measures and their sources. The formulated measures serve as clues realizing proposed potentials by multi-material design, but do not give detailed information for the design itself. The increased level of detail and the explicit reference to individual product or process characteristics such as mass, stiffness or number of assembly steps, however, support identifying fields of action, e.g. due to weak points of existing products. The measures therefore represent problem-oriented knowledge and act as a link between the strategic level (potential fields in the inner circle) and the procedural knowledge provided by the design guidelines.

The model aims to support decision-making in multi-material design and access to design guidelines. It therefore shows the allocation of general potentials and measures

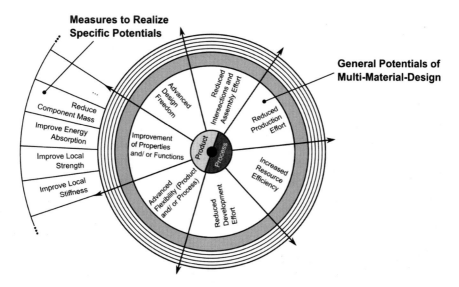

Fig. 3.4 Potential model as access level for design rules following (Kleemann 2017b)

for their realization. The semi-formal character results from the partly not explicit assignment of measures and potentials. However, the model may be understood as a checklist for assessing the suitability of individual measures for the respective design task and clarifies the breadth of the effects of multi-material design on the product, but also on the processes.

The connection between the semi-formal model of potentials and measures is established by assigning design guidelines to the defined measures. At least one design guideline is given for each measure. This enables access to the guidelines on the basis of a specific measure and the associated potential (top-down access). In addition, each design rule must be linked to the potentials and measures. This link makes it possible to understand the effect of a particular design rule with regard to the defined potentials (bottom-up analysis).

3.6   Identifying Similar Design Rules

Another feature of the knowledge management system is the content-based recommendation system. On the basis of a text similarity algorithm, it determines similar design rules and suggests them to the user.

Content-based recommendation systems are based on the content or properties of the elements. The analysed elements are then selected using a similarity criterion and suggested to the user. For unstructured, textual content, the Term Frequency-Inverse Document Frequency (TF-IDF) coding is used by default for property analysis (Jannach et al. 2010). Thus, the important words for a document are determined relative to the totality of all occurring words in all documents. Two assumptions precede this procedure according to Klahold (2009):

- Term Frequency: The more often a word occurs within a document, the more important it is.
- Inverse Document Frequency: Words that occur in many documents are less important for the document in question and its characterization

The formula for calculating this indicator is (Lops et al. 2011):

$$TF - IDF\left(t_k, d_j\right) = TF\left(t_k; d_j\right) * \log \frac{N}{n_k}$$

In addition, further approaches exist to increase the accuracy of the result (Aggarwal 2016)

- Removal of stop words: removal of words that are not element specific, but occur frequently in the respective language, such as articles, pronouns or prepositions.

- Stemming: Consolidation of variations of a term, e.g. singular and plural forms are unified.
- Extraction of Expressions: Recognition of words that are frequently used together. Their meaning often differs from that of individual terms.

3.7   Prototypic Knowledge Management System

The following section presents the prototypical implementation of the manual multi-material design. As shown in Fig. 3.5, each design guideline is defined by a short title, an explanation, pictures (good and poor example) and reference(s). To highlight the design guidelines, two figures are provided, representing unfavourable and favourable designs and providing insights for the design itself.

The design guidelines are assigned to a sensible amount of measures, stages of the development process, materials and production technologies. This allows different ways to access the information provided during the design process. Moreover, other design guidelines are suggested which are similar in terms of the assigned measures, stage of the development process, material and/or production technology. The designer is able to browse through the measures mentioned before and, in doing so, achieve a potential of multi-material design. Additionally, the user can place a request for design guidelines

Manual Multi-Material-Design

Reinforcing a structure regarding the main load-path

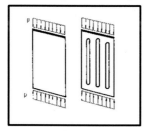
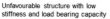
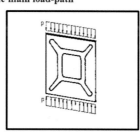

Unfavourable structure with low
stiffness and load bearing capacity

Favourable structure with higher
stiffness and load bearing capacity

Reinforcing a structure regarding the main load-paths, enlarges the local stiffness. This leads
to a higher buckling stability and, compared to a plain structure with comparable mechanical
performance, lower weight. The reinforcement may be achieved by ribs, unidirectional
composite tapes, or molded stiffeners.

References
Bernd Klein (2014), *Leichtbau-Konstruktion: Berechnungsgrundlagen und Gestaltung,*
Vieweg + Teubner Verlag.

Similar design rules
- Use of ribs for better performance
- Use of undirectional composites layer for better performance
- Use of FEA for indenfying main load-paths

Fig. 3.5 Modified "manual multi-material design"

Table 3.2 Excerpt of measures and assigned design guidelines for multi-material design

No	Measure	Design Guideline
1.1	Enlarge local stiffness	Reinforce a structure regarding the main load-paths
1.2	Enlarge local strength	Use of unidirectional composite tapes for better performance
1.3	…	…
2.1	Vary wall thicknesses	Use of composites patches for locally adjusted wall thickness
2.2	Vary material distribution	Use plastics to close surfaces
2.3	…	…

based on key words (e.g. reinforcement, stiffener, demoulding) with an optional filtering based on the potentials and phases of the development process mentioned above.

The link between the semi-formal model of potentials and measures is given by the allocation of design guidelines to the defined measures; see Table 3.2. For each measure, at least one design guideline is given. This allows an access to the guidelines based on a specific measure and the related potential (top-down access). Furthermore, the knowledge management system contains the allocation of potentials and measures to each guideline. This linking allows an understanding of the effect of a specific design guideline with regard to the potentials defined (bottom-up analysis).

Table 3.2 shows an excerpt of measures and related design guidelines. The design guidelines shown are derived from the analysed multi-material components such as automotive front-ends. Here, the use of injection moulding allows to increase the material efficacy, reduce the number of tools, reduce production steps, reduce logistics and storage, an integration of components and joining elements as well as to realize complex geometries.

The introduced knowledge management systems contain design guidelines for multi-material design. The designer is able to browse through the before-mentioned measures and potentials and, on the one hand, find suitable design guidelines. On the other hand, the system indicates which potential is affected by the single design guidelines. By now, the implementation is prototypic and available online.

3.7.1 Case Study

The following section illustrates using the manual multi-material design on a roof reinforcement of a vehicle structure as shown in Chap. 2. During the concept development, both methods for structuring the dependencies of features and properties (Kleemann et al. 2017a) and analytical design methods (Fröhlich et al. 2017) were used. In the course of concept development, design rules are also applied. Based on a functional analysis, bending, compression and torsion are identified as relevant load cases of the roof reinforcement. Bending and compression occur in a side impact when the roof reinforcement supports the b-pillars against each other. It also contributes to the global bending and torsional stiffness of the body in white. The applicability of design rules in the development of bending and torsion beams is shown (Kleemann et al. 2016). One potential of multi-material design is increasing stiffness, for example by beads and ribs. Based on the understanding of the load cases, design rules for reinforcement are applied and injection moulded ribs are selected. Cross ribs with $\pm 45°$ increase both bending and torsional stiffness of the profile and increase buckling loads as the hat profile is being supported. In further detailing, the edges of the hat profile are also crimped to increase buckling stiffness.

Furthermore, the manual multi-material design offers design rules for the injection-mouldable design of the ribs: minimum wall thicknesses and required draft angles are suggested.

3.7.2 Discussion

Design is an interplay between analysis and synthesis. The presented work provides knowledge in the form of design rules and thus primarily supports the synthesis activities. Nevertheless, further methods and tools are required for the analysis of multi-material designs. The presented work uses the potentials of multi-material design based on an analysis of current series and pre-series components in the automotive industry to structure design rules. The proposed potentials of multi-material design and the identified measures do not claim to be complete. There are several topics that have not yet been considered exhaustively, such as corporate strategic decisions such as the use of multi-material design to demonstrate technological leadership or innovative capability. In addition, the approach presented does not take costs into account sufficiently. From a unit costs point of view, multi-material design is often more expensive than conventional monolithic designs. Nevertheless, multi-material design offers advantages in production and assembly. Therefore, developers must always estimate the total costs and analyse whether the multi-material design can be amortized over the number of units produced.

References

Alexander Bischof. Developing Flexible Products for Changing Environments, Technische Universität Berlin, Diss., 2010.

Thomas Brinkmann. Handbuch Produktentwicklung mit Kunststoffen. München: Hanser, 2011.

George E. Dieter. Engineering design: A materials and processing approach. 2. ed. McGraw Hill series in mechanical engineering. New York: McGraw Hill, 1991.

K. L. Edwards, K. M. Wallace und G. Aguirre-Esponda. Designer's electronic guidebook: An engineering design guidelines database. Bd. 14. CUED/C-EDC/TR / Cambridge University Engineering Department. Cambridge: Cambridge University Engineering Department, 1993.

Klaus Ehrlenspiel, Alfons Kiewert, Udo Lindemann und Markus Mörtl. Kostengünstig Entwickeln und Konstruieren: Kostenmanagement bei der integrierten Produktentwicklung. 7. Aufl. VDI-Buch. Berlin: Springer Vieweg, 2014.

Tim Fröhlich; Sebastian Kleemann; Eiko Türck; Thomas Vietor. Multi-criteria analysis of multi-material lightweight components on a conceptual level of detail. 21ST INTERNATIONAL CONFERENCE ON ENGINEERING DESIGN, ICED17, 2017. DOI: https://doi.org/10.24355/DBBS.084-201709070942.

Armand Hatchuel und Benoit Weil. A new approach of innovative design: An introduction to C-K theory. In: The International Conference on Engineering Design (2003).

David Inkermann, Sebastian Kleemann und Thomas Vietor. Ein Potentialmodell für die Nutzung neuer Technologien in der Produktentwicklung. In: Stuttgarter Symposium für Produktentwicklung SSP 2017 (2017), p. 197–206.

Dietmar Jannach, Markus Zanker, Alexander Felfernig und Gerhard Friedrich. Recommender Systems: An Introduction. Cambridge University Press, 2010.

André Klahold. Empfehlungssysteme. Wiesbaden: Vieweg+Teubner, 2009.

S. Kleemann, E. Türck und T. Vietor. Towards knowledge based engineering for multi-material-design. In: Proceedings of International Design Conference, DESIGN DS 84 (2016), S. 2027–2036.

Sebastian Kleemann (a), Tim Fröhlich, Eiko Türck und Thomas Vietor. A Methodological Approach Towards Multi-material Design of Automotive Components. In: Procedia CIRP 60 (2017), S. 68–73.

Sebastian Kleemann (b), David Inkermann, Benjamin Bader, Eiko Türck und Thomas Vietor. A Semi-Formal Approach to Structure and Access Knowledge for Multi-Material-Design. In: Proceedings of the 21st International Conference on Engineering Design (ICED 17) DS 87-1 (2017), S. 289–298. DOI: https://doi.org/10.24355/dbbs.084-201708301114.

Bernd Klein. Leichtbau-Konstruktion: Berechnungsgrundlagen und Gestaltung. 10., überarb. u. erw. Aufl. 2013. Wiesbaden und s.l.: Springer Fachmedien Wiesbaden, 2013.

Pasquale Lops, Marco de Gemmis und Giovanni Semeraro. Contentbased Recommender Systems: State of the Art and Trends. In: Recommender Systems Handbook. Hrsg. von Francesco Ricci, Lior Rokach, Bracha Shapira und Paul B. Kantor. Boston, MA: Springer US, 2011, S. 73–105.

Richard E. Mayer und Joan K. Gallini. When is an illustration worth ten thousand words? In: Journal of Educational Psychology 82.4 (1990), S. 715–726.

Frank Nehuis, Jan Robert Ziebart, Carsten Stechert und Thomas Vietor. Use of design methodology to accelerate the development and market introduction of new lightweight steel profiles. In: ICED 2011, S. 324–330.

Mark L. Nowack. Design guideline support for manufacturability. Diss. University of Cambridge, 1997.

Karlheinz Roth. Konstruieren mit Konstruktionskatalogen: Band 1: Konstruktionslehre.3. Auflage, erweitert und neu gestaltet. Berlin: Springer, 2000.

Filippo A. Salustri, Nathan L. Eng und Janaka S. Weerasinghe. Visualizing information in the early stages of engineering design. In: Computer-Aided Design and Applications 5.5 (2008), S. 697–714.

Helmut Schürmann. Konstruieren mit Faser-Kunststoff-Verbunden. 2.,übearbeitete und erweiterte Auflage. VDI-Buch. Berlin, Heidelberg: Springer-Verlag Berlin Heidelberg, 2007.

Snowden, Dave (2002). "Complex Acts of Knowing – Paradox and Descriptive Self Awareness". Journal of Knowledge Management. 6 (2): 100–111. CiteSeerX 10.1.1.126.4537. doi:https://doi.org/10.1108/13673270210424639.

Sandor Vajna. WISSENSMANAGEMENT IN DER PRODUKTENTWICKLUNG. In: DFX 2001. Hrsg. von H. Meerkamm. 11.-12.10.2001.

Michael James van Wie. Designing product architecture: A systematic method. 2002.

Christian Weber. Looking at "DFX" and "Product Maturity" from the Perspective of a New Approach to Modelling Product and Product Development Processes. In: The Future of Product Development. 2007.

Jan Robert Ziebart. Ein konstruktionsmethodischer Ansatz zur Funktionsintegration: Braunschweig, Techn. Univ., Diss., 2012. München: Dr. Hut, 2012.

Anja Jäschke and Ulrich Dajek. Dachrahmen in Hybridbauweise. In: Sonderdruck aus VDI-Tagungsband Nr. 4260. p. 25-45. Düsseldorf: VDI Verlag GmbH, 2004.

LANXESS Deutschland GmbH (2010, 14. June). Frontend in Hybridtechnik mit Organoblech. LANXESS. URL: https://techcenter.lanxess.com/scp/emea/de/docguard/LANXESS_Durethan_BKV30H2.0EF_-_Audi_-_Frontend_Tepex_-_CS_TI_2010-002_DE.pdf?docId=14316664.

Maik Gude, Holger Lieberwirth, Gerson Meschut, Michael F. Zäh (Eds.). FOREL-Studie - Chancen und Herausforderungen im ressourceneffizienten Leichtbau für die Elektromobilität. 2015.

Charu C. Aggarwal. Recommender Systems. The Textbook. Springer Cham. 2016.

Levers of Cyber Physical Production Systems for Multi-Material Body Parts Manufacturing

4

Sebastian Gellrich, Christoph Herrmann and Sebastian Thiede

Abstract

The manufacturing of multi-material components for the automotive industry adds additional process steps to the traditional process chain while also employing new manufacturing technologies. Both factors usually lead to higher production costs as well as higher energy demands and environmental impacts. Cyber physical production systems (CPPS) are a promising approach to support the design of multi-material components as well as the dimensioning and operation of the production processes. In the following article, the levers of CPPS for the manufacturing of multi-material components are discussed and illustrated within an automated calculation of component-specific energy consumption.

4.1 Introduction

The value chain of multi-material components encompasses several converging process steps—from different raw materials, e.g. plastic granulate, over pre-products like organosheets to their hybridization, e.g. thermoforming. This chain results in complex

S. Gellrich (✉) · C. Herrmann · S. Thiede
Institute of Machine Tools and Production Technology (IWF), Technische Universität
Braunschweig, Braunschweig, Deutschland
e-mail: s.gellrich@tu-braunschweig.de

C. Herrmann
e-mail: c.herrmann@tu-braunschweig.de
S. Thiede
e-mail: s.thiede@tu-braunschweig.de

© Springer-Verlag GmbH Germany, part of Springer Nature 2023   55
T. Vietor (ed.), *Life Cycle Design & Engineering of Lightweight Multi-Material
Automotive Body Parts*, Zukunftstechnologien für den multifunktionalen Leichtbau,
https://doi.org/10.1007/978-3-662-65273-2_4

interdependencies, a high number of influencing factors and is often related with high manufacturing cost and additional energy consumption, caused by complex processes or pre-heating activities. In order to understand and control this new complexity, the analysis of manufacturing data appears to be a promising approach. As an example, the derivation of part-specific energy data especially for new production technologies is investigated. Data analytics approaches can support in overcoming the lack of reliable life cycle inventory (LCI) data to enable an appropriate assessment of the components environmental impact. Cyber physical production systems (CPPS) may serve as a viable concept for efficiently supporting the design and manufacturing of multi-material components. In addition, the CPPS framework supports in structuring the technological functionalities that are necessary for an adequate analysis of manufacturing data.

4.2   Cyber Physical Production Systems

CPPS extends common IT solutions in industry, like supervisory control and data acquisition (SCADA), with specific new and tailored functionalities on machine to factory level (Thiede 2018). In general, cyber physical systems are defined as "systems of collaborating computational entities which are in intensive connection with the surrounding physical world and its on-going processes, providing and using, at the same time, data-accessing and data-processing services" (Kang et al. 2016). A CPPS framework with its four subsystems (I–IV) and their interconnections for information exchange is shown in Fig. 4.1. The framework shows that within a CPPS data of the physical world (I), e.g.

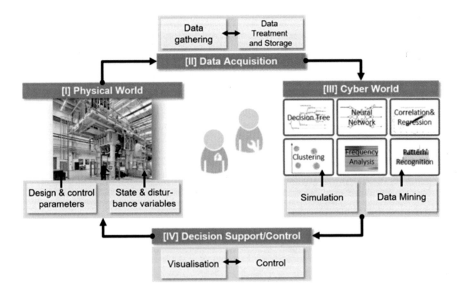

Fig. 4.1 General framework of a CPPS. (adapted from Thiede et al. 2016, picture: OHLF/Wecke)

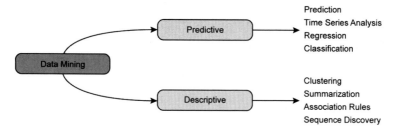

Fig. 4.2 Overview of data mining tasks, adapted from (Dunham 2003)

design and control parameters, as well as external influencing factors, are continuously acquired (II) if possible in high temporal and spatial resolution. Subsequently, the data is assessed within the cyber world (III) by data mining and/or simulation models. The modelling results are fed back into the physical world as decision support or machine control (IV). Within this framework, the human stays in focus, e.g. as expert for model building and validation or being assisted in process control (Thiede 2018). As depicted above, information from manufacturing data is gained through simulation or data mining, whereby the focus of this article is on data mining. In the following, tasks of data mining and its role in supporting the engineering of multi-material components are discussed.

Larose et al. (2014) define data mining as "the process of discovering useful patterns and trends in large data sets" (Larose et al. 2014). Models derived by data mining can be divided into predictive and descriptive models. The former makes predictions of target variables based on known historical data; this is why they are also called supervised learning methods. The latter finds patterns and relationships in the data without knowing the relationships, explaining why they are called unsupervised learning methods. As shown in Fig. 4.2, representatives of predictive data mining are prediction, time series analysis, regression and classification. Clustering, characterization, association rules and sequence discovery are associated with descriptive data mining.

The data mining approach of prediction focuses on predicting future states of target variables based on other historical and actual data, while regression and classification predicts current states. In time series analysis, an attribute change of value is analysed over the time, often with a fixed sampling rate. In regression tasks, a real-valued target variable is obtained, by mapping a function (e.g. logistic) on other known historic data that minimizes an error function. Closely related to regression, classification uses known historic data, to map input data into target groups or classes. Likewise, similar to classification, the unsupervised method of clustering searches also for classes, but not predefined ones. The clusters are built by analysis of similarity or proximity of attributes. By applying characterization (also called summarization), representative simple descriptions of the data are derived. The task of association rules mines relationships among data, which are described and can be investigated by different metrics like support, confidence and lift. Closely related to association rules, but based on time, sequence discovery is applied for revealing sequential patterns in data (Dunham 2003). All the before-mentioned data mining tasks are viable candidates to be leveraged for an informed engineering of multi-material structures, whereby the exemplary application in Sect. 4.4 will have a focus on classification.

4.3 CPPS-based Architecture

4.3.1 Levers of Data-Based Modelling for the Manufacturing of Multi-Material Body Parts

For multi-material components, e.g. using carbon fibres, the process–structure–property relationships are complex and difficult to control. In addition, energy-intensive processes are often required. Thus, the manufacturing of multi-material components leads in many cases to increased manufacturing cost and potential environmental impacts. For better competitiveness, data-based approaches can support both component design and process design and operation, providing a valuable contribution to eco-efficient multi-material components. Data-based methods can be used in various manufacturing disciplines to increase productivity. Widely discussed applications are, for example, the prediction of machine failures and the efficient planning of maintenance measures to avoid unplanned downtimes (e.g. Neef et al. 2018) and root-cause analysis in quality management (e.g. Gellrich et al. 2019a). Data-based methods can be applied on several manufacturing disciplines dealing with multi-material components as shown in Fig. 4.3.

Within **process design** (1), parameters influencing the process are identified and modelled in order to optimize the overall process based on the derived models. The control of a specific machine behavior, e.g. by means of a machine state recognition to efficiently

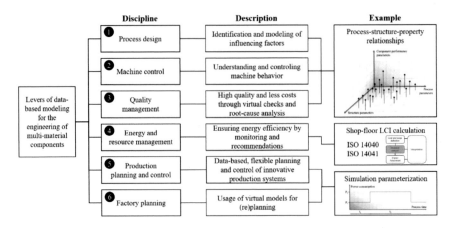

Fig. 4.3 Levers of data-based modelling for the engineering of multi-material components

switch to energy-saving states, is part of **machine control** (2). Through **virtual quality management** (3), complex physical quality inspections can be reduced or virtualized, and valuable insights on process–structure–property relationships can be drawn. This enables robust process design as well as process-related robust component design. For example, in injection overmoulding of short FRTP, possible relationships could be the dependencies between process and structure parameters, e.g. melt temperature and flow rate on the resulting fibre orientations, as well as between structure parameters and component performance properties, e.g. fibre orientation on stiffness. In the field of **energy and resource management** (4), energy monitoring systems could be enriched by prediction and anomaly detection functionalities. The automated online prediction of LCI data for production processes can also be based on data mining approaches. This real-time information can be used to increase the awareness for environmental impacts of the current operation mode among shop floor workers and management. Additionally, this information delivers a basis for a more realistic environmental assessment in life cycle-oriented product design. An exemplary application for an automated calculation of LCI data for a multi-material component production process is shown in Sect. 4.4. Finally, data-based methods can also be deployed for efficient production **planning and control** (5) as well as **factory planning** (6). Especially for innovative production systems (e.g. matrix-production), data-based methods can be leveraged for fast and accurate insights. The prepared and condensed manufacturing data can also be used to parameterize process simulation models.

4.3.2 CPPS-Based Architectural Approach

In order to efficiently apply data-based modelling for multi-material structures, a sufficient data acquisition and treatment infrastructure is required. In the following, an architectural approach is depicted, which is exemplified by the data infrastructure of the Open Hybrid LabFactory. At the research campus, a huge variety of innovative production technologies covering the complete value chain of multi-material components can be found in the technical centre, e.g. textile machines for organic sheets, low-pressure die casting, hybrid injection moulding and a 2,500 t hybrid forming press. In order to support a comprehensive understanding of these new technologies, a digitalization approach is pursued. The approach is based on the framework of CPPS (see Sect. 4.2 **Fehler! Verweisquelle konnte nicht gefunden werden.**). The phases serve as layers of the architectural approach. The CPPS-based data infrastructure of the Open Hybrid LabFactory is shown in Fig. 4.4. The figure highlights an energy monitoring use case on process and factory level. The layers are described in the following.

[I] **Physical World:** As depicted in Fig. 4.4, manufacturing data is acquired from shop floor via programmable logic controllers (PLCs), decentralized sensors and energy meters. Further valuable data for analysis tasks, like ambient data, can be accessed through low-cost wireless sensors that are capable of modern communication

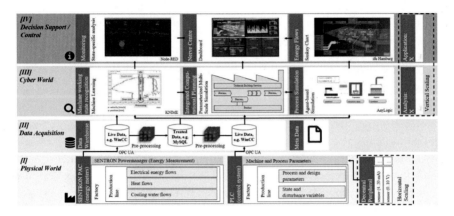

Fig. 4.4 CPPS-based data architecture at the Open Hybrid LabFactory

protocols like MQTT (XDK Cross Domain Development Kit n.d.). At the Open Hybrid LabFactory, the data acquisition infrastructure for energy data is fully automated and exhaustive. Each machine of the technical centre is equipped with at least one 7 km PAC energy meter (3100, 3200 or 4200) (SENTRON Messgeräte und Energiemonitoring n.d.). Dependent on the devices version, a huge number of electrical parameters are measured, e.g. current, voltage, power, energy, frequency and total harmonic distortion. In total, 33 energy meters are installed for production machines, 12 for laboratories and 28 for technical building services. The measured data is collected, stored and can be monitored on a SCADA system for energy data. The SENTRON powermanager (SENTRON Messgeräte und Energiemonitoring n.d.) offers historical data access, live monitoring as well as a gateway functionality (OPC server) for further data-processing tasks, e.g. non-proprietary software tools for data storage and analysis (see layer [II]). For PLC access, the SCADA software SIMATIC WinCC V7 is used as a gateway (SIMATIC WinCC V7 n.d.).

[II] Data Acquisition: Various applications emanating from different engineering disciplines deliver requirements concerning an efficient data access. For example, in life cycle engineering, real-time life cycle assessments may be performed, which requires a low-latency data flow (Cerdas et al. 2017). Other applications like data-driven root-cause analysis of manufacturing data on resulting component performance, e.g. bending test, requires an efficient access to stored data. The capability of an IT infrastructure to host flexible applications is widely discussed in terms of the service-oriented architecture. Consequently, the data acquisition layer has to serve the requirements of (near) real-time and historical data access. The lambda architecture is one approach to satisfy these requirements (Marz et al. 2015). The architecture consists of a speed, batch and serving layer. The speed layer handles real-time data flows. The data is stored temporarily and is used to parameterize precomputed models. In Fig. 4.4, the speed layer is represented by the live data bases and their direct links to the cyber world layer. The architecture batch and a serving layer are for historical data access, represented by the data base labelled with treated data in Fig. 4.4. Those layers have a higher latency than the speed layer, whereby pre-processed data, e.g. a machines working state, is stored persistently in the batch layer and tailored calculations on the historical data are stored in the serving layer in a service-oriented manner. The acquired manufacturing data can be enriched via metadata such as performance results of a component. The lambda architecture and data pre-processing, e.g. filtering of irrelevant raw data, ensures a flexible and cost-efficient data acquisition layer.

[III] **Cyber World:** Models build through data mining and simulation represent the cyber world layer. The output of the cyber world models, e.g. decision on a good or bad product quality, is forwarded to the decision support and control layer. Real-time and/ or historical manufacturing data serves as input for model deployment and training. Because of changing production conditions, the validity of the models needs to be verified at certain intervals.

[IV] **Decision Support and Control:** The design of the decision support and control layer strongly depends on the application scenario. The fully automated CPPS loop is accomplished through direct write access to the production machines. Decision support applications have the human in focus. The decision support is delivered via a human–machine interface. This can be done directly on shop floor by devices like wearables, e.g. smart watches, or in a remote manner, for example via a production control centre. At the Open Hybrid LabFactory a video wall at the Life Cycle Design & Engineering Lab serves as the monitoring platform for the manufacturing activities in the technical centre.

4.4   Exemplary Use Case: Deriving Product-specific Energy Consumptions through Data-based Modelling

The following section demonstrates the applicability of the developed approach based on use case. For the manufacturing of two multi-material components on a forming press at the Open Hybrid LabFactory, the actual energy consumption is calculated through data-based modelling. Firstly, a rib structure made of GFRP is extruded onto a hybridized u-profile (see also Gellrich et al. 2019b). Secondly, an FRP tray is produced by thermoforming, and the respective energy consumption is determined on the basis of an automated machine state recognition. For this purpose, the CPPS-based architecture introduced in the previous section serves as a framework.

Physical World—Hybrid Forming Press

The extrusion and thermoforming are carried out on a 2,500 t metal forming press (see Fig. 4.5). The multi-functional press is capable of different process variants like cold forming, warm forming, sheet moulding compound (SMC) and resin transfer moulding (RTM) and is characterized by a combination of high pressing force and high speed. For the extrusion process, the holding time and temperature are varied in a total of 18 test runs. The

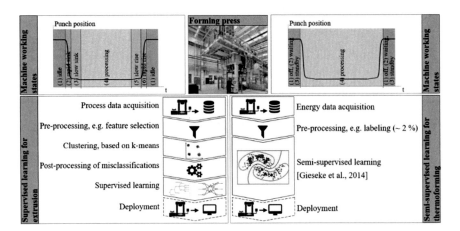

Fig. 4.5 Machine states and data mining workflows for exemplary use cases (picture: OHLF/ Wecke)

process is characterized by six machine states. In idle mode, which is defined as machine state 1, the punch is in its highest position. The working stroke begins with a rapid traverse (rapid sink of the punch—machine state 2), which is followed by slow sink (machine state 3) when entering the die. During processing (machine state 4), the punch is held constant. When processing is finished, the retraction starts with a slow rise of the punch (machine state 5), until it changes into rapid rise / retraction (machine state 6) and finally idle mode again (machine state 1). For thermoforming, data of only six strokes is acquired. The working states to be recognized are reduced to off, standby, idle and processing.

Data Acquisition—Process and Energy Data

For extrusion, the machine learning algorithm is trained on process data, comprising 48 features with a sampling rate of 50 ms. A correlation analysis yields four independent features which are chosen for model training (press force [kN], speed [mm/s], supply pressure [bar] and cylinder force [kN]). For deriving the part-specific energy consumption, energy data needs to be acquired, too. In order to reduce this effort in matching two data sources, the second use case of thermoforming is trained directly on energy data, i.e. active power [W], as input. The energy data for both process variants is acquired with a time resolution of 1 s.

Cyber World—Machine States Recognition based on Supervised and Semi-Supervised Learning

Two different modelling workflows and approaches based on classification are deployed for machine working states recognition. The working states are subsequently used as classes for actual energy consumption mapping, delivering a state-specific energy consumption per manufactured component. The first approach that is applied on extrusion

Table 4.1 Comparison of properties of supervised and semi-supervised learning for machine state recognition

Approach Properties	Supervised Learning for Extrusion	Semi-Supervised Learning for Thermoforming
Data	Process data	Energy data
Sampling rate	~50 ms	~1 s
Modelling steps	2 (Clustering; Classification)	1 (Label propagation)
Manual data-processing effort	High (Data post-processing)	Low (2,1% labelled samples)
Modelling performance	High (99.7%)	High (97.3%)
Energy KPI calculation	Requires matching with energy data	Directly computable

is a two-step modelling process with an initial clustering and subsequent supervised learning. The second approach, applied for thermoforming, only requires a single modelling stage through the application of semi-supervised learning. The properties of both approaches are compared in Table 4.1.

Supervised Learning based on Process Data for Extrusion

Supervised learning methods require historic data with known target variables (here: machine states). As the data acquired lacks the labelled target variable and in order to reduce the manual effort of labelling thousands of data points, clustering in terms of a k-means is applied. The method supports in a rough pre-labelling of the data set. By iteratively increasing the number of clusters and plotting the clustering results, an assessment whether all targeted machine states are successfully separated or not can be performed ($k = 8$, yielded all six required machine states for the extrusion use case). The clusters found for $k = 6$ and 7 describe idle mode and are pooled together. Subsequently to clustering, the mislabelled data points are manually edited to yield an ideal labelled data set. For the supervised learning task, four models are considered (decision tree, random forest, support vector machine and probabilistic neural network). Through this comparison, the best fitting method for classifying the machine states based on the performance metrics precision and recall can be identified. Random forest proved to be the best method for the acquired input data reaching an accuracy of 99.7%. In contrast, the neural network approach performed worst, by having zero recall/precision for the slow sink state (machine state 3). The classification for this class is improved by applying an under-sampling on the training data, achieving a recall of 0.46 and precision of 0.83.

Semi-supervised learning based on energy data for thermoforming

In order to address the weaknesses of the first approach, i.e. two modelling steps, high manual effort in data pre-processing / postprocessing as well as matching with energy

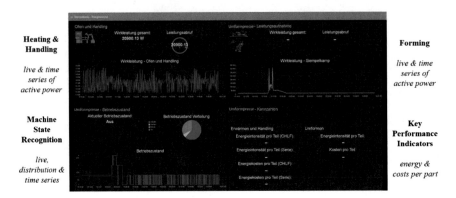

Fig. 4.6 Forming press energy dashboard at the Open Hybrid LabFactory (picture: Siempelkamp)

data for the calculation of part-specific energy consumption, a second approach for thermoforming of an FRP tray is tested. As data input, the active power of the process is selected in order to avoid the matching of two data sources. To save both, a modelling step and manual effort, the method of semi-supervised learning, in terms of label propagation, is used. Semi-supervised learning approaches require only a small amount of labelled data. However, unlabelled data is considered within training, e.g. via density functions, and thus increases the robustness of the model (Gieseke et al. 2014). In the example data set, only 2.1% of the data points were labelled manually with the four machine states. The trained model yielded a similarly high accuracy of 97.3%. On the basis of the classified data, the part-specific energy consumption during the processing state can be easily calculated.

Decision Support—Dashboard for Advanced Process Monitoring
For closing the CPPS loop, a prototypical decision support system has been developed. As shown in Fig. 4.6, the dashboard offers different functionalities for process monitoring: Live energy data and time series data of the relevant process chain (here: heating and handling of pre-products for the forming process), the live machine state and its historic distribution, and finally, energy and cost key performance indicators, e.g. energy per part. The *dashboard* is implemented in the open-source software Node-RED (Node-RED n.d.).

4.5   Summary

The manufacturing of multi-material components is faced with a complex converging process chain and new production technologies. The approach of CPPS offers viable levers to control this increased complexity in manufacturing and can also add value to the engineering of multi-material components. This chapter has illustrated different levers of CPPS and provides a concept based on the CPPS framework for an adequate data architecture to meet the presented levers. The exemplary use case shows how this architecture can be used to automatically determine part-specific energy indicators for the manufacturing of multi-material components.

References

Cerdas F., Thiede S., Juraschek M., Turetskyy A. & Herrmann C. (2017). Shop-floor Life Cycle Assessment. Procedia CIRP, 61, 393–398. doi: https://doi.org/10.1016/j.procir.2016.11.178

Dunham M. H. (2003). Data Mining: Introductory and Advanced Topics. Pearson Education. ISBN:978–0–13–088892–1

Gellrich S., Beganovic T., Mattheus A., Herrmann C. & Thiede S. (2019a). Feature Selection Based on Visual Analytics for Quality Prediction in Aluminium Die Casting. IEEE 17th International Conference on Industrial Informatics (INDIN). 66–72. doi: https://doi.org/10.1109/INDIN41052.2019.8972093

Gellrich S., Filz M.-A., Wölper J., Herrmann C. & Thiede S. (2019b). DATA MINING APPLICATIONS IN MANUFACTURING OF LIGHTWEIGHT STRUCTURES. In: Dröder K, Vietor T, eds. Faszination Hybrider Leichtbau / Technologies for Economical and Functional Lightweight Design. Berlin, Heidelberg: Springer Vieweg; 15–27. doi: https://doi.org/10.1007/978-3-662-58206-0_2

Gieseke F., Airola A., Pahikkala T. & Kramer O. (2014). Fast and simple gradient-based optimization for semi-supervised support vector machines. Neurocomputing, 123, 23–32. doi: https://doi.org/10.1016/j.neucom.2012.12.056

Kang H. S., et al. (2016). Smart manufacturing: Past research, present findings, and future directions. Int. J. Precis. Eng. Manuf. - Green Technol., 3, 1, 111–128. doi: https://doi.org/10.1007/s40684-016-0015-5

Larose D. T. & Larose C. D. (2014). Discovering Knowledge in Data: An Introduction to Data Mining, Wiley Series on Methods and Applications in Data Mining. doi: https://doi.org/10.1002/9781118874059.ch1

Marz N. & Warren J. (2015). Big Data: Principles and best practices of scalable real-time data systems. Manning Publications. ISBN:9781617290343

Neef B., Bartels J. & Thiede S. (2018). Tool Wear and Surface Quality Monitoring Using High Frequency CNC Machine Tool Current Signature. IEEE 16th International Conference on Industrial Informatics (INDIN), 1045–1050. doi: https://doi.org/10.1109/INDIN.2018.8472037

Node-RED: Low-code programming for event-driven applications. Retrieved from https://nodered.org/ (last visited 04.08.2020)

SENTRON Messgeräte und Energiemonitoring. Retrieved from https://new.siemens.com/global/de/produkte/energie/niederspannung/komponenten/sentron-messgeraete-und-energiemonitoring.html (last visited 04.08.2020)

SIMATIC WinCC V7. Retrieved from https://new.siemens.com/global/de/produkte/automatisierung/industrie-software/automatisierungs-software/scada/simatic-wincc-v7.html (last visited 04.08.2020)

Thiede S. (2018). Environmental Sustainability of Cyber Physical Production Systems. Procedia CIRP, 69, 644–649. doi: https://doi.org/10.1016/j.procir.2017.11.124

Thiede S., Juraschek M. & Herrmann C. (2016). Implementing Cyber-physical Production Systems in Learning Factories. Procedia CIRP, 54, 7–12. doi: https://doi.org/10.1016/j.procir.2016.04.098

XDK Cross Domain Development Kit. Retrieved from https://xdk.bosch-connectivity.com/de/overview (last visited 04.08.2020)

Modeling and Simulation of New Manufacturing Processes for Multi-Material Lightweight Body Parts to Estimate Environmental Impacts

5

Antal Dér, Christopher Schmidt, Christoph Herrmann and Sebastian Thiede

Abstract

Recent years introduced process and material innovations in the design and manufacturing of lightweight body parts. Lightweight materials and new manufacturing processes often carry a higher environmental burden in earlier life cycle stages. The prospective life cycle evaluation of newly developed manufacturing processes and related production systems remains to this day a challenging task. Against this background, this chapter introduces a modeling and simulation approach for determining the potential environmental impacts of new manufacturing processes and production systems for multi-material lightweight body parts.

A. Dér (✉) · C. Schmidt · C. Herrmann · S. Thiede
Institute of Machine Tools and Production Technology (IWF), Technische Universität Braunschweig, Braunschweig, Deutschland
e-mail: a.der@tu-braunschweig.de

C. Schmidt
e-mail: christopher.schmidt@tu-braunschweig.de

C. Herrmann
e-mail: c.herrmann@tu-braunschweig.de
S. Thiede
e-mail: s.thiede@tu-braunschweig.de

© Springer-Verlag GmbH Germany, part of Springer Nature 2023
T. Vietor (ed.), *Life Cycle Design & Engineering of Lightweight Multi-Material Automotive Body Parts*, Zukunftstechnologien für den multifunktionalen Leichtbau,
https://doi.org/10.1007/978-3-662-65273-2_5

5.1   Introduction

Vehicle manufacturers have been increasingly developing and utilizing multi-material lightweight body parts to reduce fuel consumption. Multi-material lightweight body parts, as referred to in this chapter, consist of fibre-reinforced polymers (FRP). They are employed in vehicle body structures in addition to either conventional steel structures or as a replacement for conventional steel-based designs. While FRPs do help to reduce the vehicle's weight, hence reduce fuel consumption and corresponding CO_2 emissions during the use stage of conventional vehicles with an internal combustion engine; they tend to create a higher environmental burden during material production and parts manufacturing than their counterparts (Herrmann et al. 2018). Fig. 5.1 illustrates this aspect by highlighting the cradle-to-gate impacts from the manufacturing stage. The figure illustrates the environmental impact of the subsequent processing steps in a qualitative step chart. The additional environmental impact of lightweight body parts in comparison to steel parts can be explained as follows. First, the embodied energy of lightweight materials tends to be higher than that of steel, today's mainstream engineering material (Duflou et al. 2012). Carbon fibre-reinforced plastics as an example have an embodied energy that is over five times higher than steel (Herrmann et al. 2018). Furthermore, process chains for multi-material lightweight body parts encompass new and altered manufacturing processes with higher lead times (due to curing times) and energy demands (due to thermal process steps) in comparison to well-established steel-based automotive process chains (Fleischer et al. 2018). Together, this leads to an increase of the cradle-to-gate environmental impact of manufacturing of lightweight body parts. Consequently, a high material efficiency and energy-efficient processes may be two levers to limit excess cradle-to-gate environmental impact of lightweight body parts.

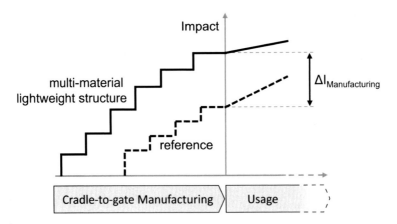

Fig. 5.1 Cradle-to-gate environmental impacts of the manufacturing of multi-material lightweight body parts in comparison to steel-based reference body parts, inspired from (Herrmann et al. 2018; Kaluza et al. 2017; Schönemann et al. 2016))

In order to reduce the environmental impact over the entire life cycle of lightweight vehicles, it is paramount to understand the environmental impact of producing lightweight body parts and how it may be reduced. To this end, product and production planners should be able to effectively assess the environmental impacts of producing lightweight body parts in concept design phases. A challenge towards this early assessment is however the lack of empirical data. This chapter focuses on estimating the potential environmental impacts with a focus on energy intensity of new manufacturing processes for multi-material lightweight body parts. To this end, a modelling and simulation approach has been developed that integrates bottom-up process models into a process chain model and predicts the energy demand of the manufacturing of lightweight body parts. First, an overview about new process chains and manufacturing processes is given. This is followed by a brief review of relevant research approaches and the presentation of the proposed concept. A case study rounds up the chapter by demonstrating the implementation and functionality of the method.

5.2   Manufacturing Processes for Multi-Material Lightweight Body Parts

Established high-volume manufacturing processes for automotive structural components are based on sheet metal processing (Ingarao et al. 2011). FRP-based multi-material structures exploit their full lightweight potential in load-path optimized structures, which calls for the development of new manufacturing processes that fulfil the requirements of high-volume production (Buschhoff et al. 2016). Fig. 5.2 illustrates an overview of an FRP-based process chain for manufacturing automotive structural components. FRP process chains encompass textile processes for semi-finished parts processing and a number

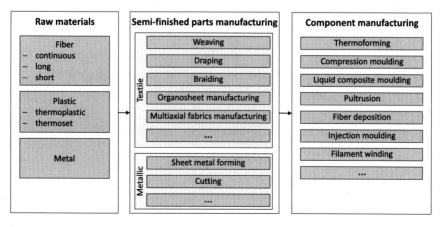

Fig. 5.2 Manufacturing processes for multi-material lightweight body parts, compiled from (Dröder et al. 2014; Fleischer et al. 2018)

of processes depending on part geometry and matrix material for final part production (Dröder et al. 2014). Hybrid process chains emerge from the combination of intrinsic metal and FRP processing and show promising results towards being a competitive alternative to traditional parts manufacturing (Fleischer et al. 2018).

High production volumes in the automotive industry and resulting low cycle times increase the pressure on the development of suitable manufacturing processes for multi-material structures. In this regard, a major challenge will be to qualify current lab-scale processes with low MRL for high-volume production. Another challenge is to integrate manufacturing processes with high MRL into the established automotive process chain and factory environment, while justifying the advantages from a life cycle perspective (Fleischer et al. 2018; Herrmann et al. 2018).

5.3 Determining Environmental Impacts of New Manufacturing Processes via Modelling and Simulation

Modelling and simulation have been accepted over the previous years as a suitable approach for investigating the energy demands and increasing the energy efficiency of manufacturing systems (Garwood et al. 2018). Modelling and simulation approaches of recent years address different scales of production systems, ranging from process and machine, e.g. (Abele et al. 2015) over to process chains including technical building services, e.g. (Herrmann et al. 2011) and holistic multiscale production system simulations including all levels of a production system (Schönemann et al. 2019).

Herrmann et al. (2011) developed an energy-oriented simulation approach that is able to consider all relevant energy flows of factory systems and their dynamics. The approach acts as a decision support for the systematic improvement of for manufacturing systems. Schönemann et al. introduce a multi-level modelling and simulation approach in the context of manufacturing lightweight body parts. They combine multiple models on different scales (e.g. product, process, process chain and factory building) to simulate the accurate energy demand of manufactured products (Schönemann et al. 2016).

The characterization of the energy demand based on empirical models has also been investigated for different material removal processes and injection moulding (Li 2015). The focus of previous approaches, however, lies rather in predicting energy demands for the planning and optimized operation of a manufacturing system. In order to support the collection of LCI data of manufacturing unit processes, Kellens et al. (2012) provide a methodology that includes a time, power, consumables and emissions study. The methodology distinguishes between different machine states (e.g. standby and processing mode) and allocates state-based energy, consumables and emissions data. The methodology suggests the measurement or estimation of related data. However, the lack of empirical data for new and yet-to-be-developed manufacturing processes for multi-material lightweight body parts is an obstacle towards directly applying this approach.

5.4   Approach and Implementation for Multi-Material Lightweight Body Parts

Simulation has proven to be an appropriate method to evaluate and analyse potential problem shifting and dynamic interrelationships within manufacturing systems. It is a preferred method for cases where empirical data is scarce, the studied system is complex and includes dynamic characteristics between manufacturing system elements. Furthermore, simulation can be easily applied to assess what-if scenarios that would not be economically feasible within real production environments. (Schönemann et al. 2019).

The modelling approach, shown in Fig. 5.3, proposes the combination of energy-oriented bottom-up machine models in a generic process chain modelling environment. The component-based machine models calculate the state-based energy demand of the main machine components, such as drives, tempering units and hydraulic components. Several machine models constitute the process-modelling library, which provides the simulation

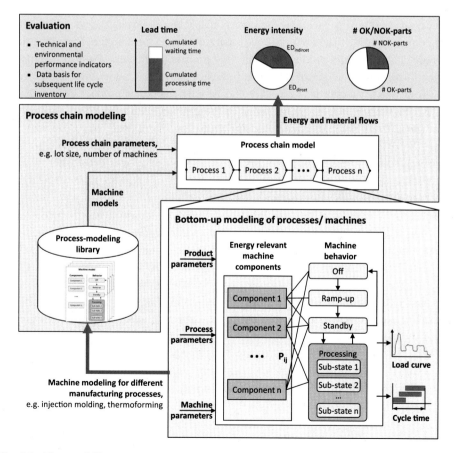

Fig. 5.3 The modelling approach, inspired from (Kaluza et al. 2020)

modeller with building blocks for the process chain model. The generic process chain model represents a modelling environment for flexibly setting up process chains and embedding them in a factory environment.

5.5   Bottom-up Modelling of Processes

The modelling of different machines follows the same generic logic regardless of the type of machine. Following states are modelled similar to (Thiede 2012): off, ramp-up, standby and processing. The processing state is machine-specific and includes further sub-states to accurately model the machine behaviour. The output of a machine model is its load curve, cycle time and material utilization. All output parameters are successively used in the process chain simulation model to calculate the product's energy intensity.

For product concepts in early design stages, the simulation approach assumes that there is neither empirical data available in form of electrical measurements nor data for other energy forms to parametrize the machine models. In these cases, the machine's electrical load curve first needs to be approximated using physical laws and a set of assumptions regarding machine and process parameters. Processing times and electricity power are calculated with product, machine and process parameters. Product parameters describe those relevant characteristics that influence the direct energy input for processing such as mass, specific heat capacity and geometry. Further product characteristics that describe the mechanical performance of the product such as ductility, strength or fibre orientation are not part of the modelling approach. Process and machine parameters provide the basis for calculating the electricity input for the machine components. The calculations follow a backward logic: first defining the energy required for the production process itself and secondly calculating each component's power demand considering the efficiency of machine components.

For product concepts in later design stages, which are already in prototype or single production, the modelling approach is extended to integrating electrical measurements and other measured energy demands such as compressed air into the machine models. In these cases, the measured energy load curves are matched with the modelled machine states. In order to explore the impact of cutting waste in forming and separating processes and sprue waste in primary shaping processes, material efficiency is a variable in the machine models. Additionally, the quality rate of the manufacturing processes can be adjusted as well.

5.6   Process Chain Modelling Environment

Summing up the energy demand of isolated machine models for manufacturing a product is not sufficient for predicting the total energy intensity of production. Isolated machine models do calculate the direct energy demand of one product. They neglect however indirect energy demands for waiting times, machine ramp-ups and the embodied energy in scrap parts. Combining machine models for simulating component manufacturing allows for considering dynamic interrelationships along the process chain, e.g. the effect of the quality rate on the indirect energy demand. The dynamics of a process chain materialize into different utilization of machines, lead times and peak loads. Therefore, breaking down the total energy intensity of one product to direct and indirect energy demands can help to better understand each machine's contribution to the energy intensity and consequently environmental impact.

The process chain modelling environment ensures that machine models from the process-modelling library can be integrated into a process chain. Necessary inputs are therefore the order of the product's processing sequence and the lot size. The output is the product's energy intensity EI, which is calculated using the following equation:

$$EI = \sum_{1}^{n}(ED_{direct} + ED_{indirect})$$

The model logic distinguishes between the direct energy demand ED_{direct} and the indirect energy demand $EI_{indirect}$. Both direct and indirect energy demands are summed up over all process steps to form the product's energy intensity. The distinction between direct and indirect energy demands stems from the definitions provided by (Posselt et al. 2014; Seow et al. 2011). In this modelling approach, the direct energy demand arises from the processing stage of the machine while the energy demand from all other stages accounts for the indirect energy demand. The direct energy demand of scrap parts is equally divided among good parts and counted as indirect energy demand. Consequently, lower quality rates would increase the indirect energy demand of good parts.

5.7 Case Study—Modelling and Simulation of Manufacturing Processes at the Open Hybrid LabFactory

A case study was carried out at the Open Hybrid LabFactory (OHLF) to demonstrate the functionality of the proposed concept. The OHLF covers a wide range of a multi-material lightweight automotive component's value chain, ranging from textile processes for semi-finished parts manufacturing to final parts manufacturing with injection moulding and thermoforming. Fig. 5.4 shows an exemplary process chain of a lightweight FRP tunnel with a rib structure. The manufacturing of the tunnel consists of three manufacturing steps. First, the glass fibre-reinforced organosheet is heated up in an infrared oven. Afterwards, it is thermoformed in a hydraulic press. In the last step, a rib structure is formed on the inner side of the tunnel via injection moulding.

The modelling environment for both the machine modelling and process chain modelling, as shown in Fig. 5.5, is AnyLogic. The models are however connected with an Excel interface that both provides both the input data and saves the results from the

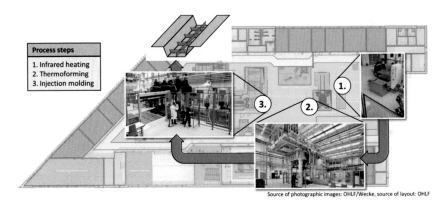

Fig. 5.4 Exemplary process chain at the Open Hybrid LabFactory

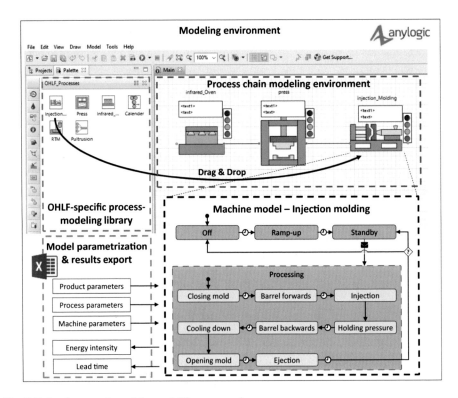

Fig. 5.5 Implementation of the modelling approach

simulation runs. The model takes advantage of the combination of the agent-based modelling approach with the discrete event modelling approach. Products represent agents that enter the discrete event process chain according to a pull request from the first machine. The machine agents have a standardized interface for the communication with the product agents. The interface consists of an input and output port that ensures the machine's connectivity with other objects in the process chain model. After a product agent arrives at the machine object, it sends a message to the corresponding machine agent that signalizes product arrival and the machine switches from its standby to processing state. The machine executes its processing steps according to its statechart and calculates component-wise the state-based electric power demands. Following components were identified as relevant and modelled on the example of the injection moulding machine: electric motor for powering the hydraulic pump, cooling unit for the mould and hydraulic system and the heating system for the barrel.

The process chain modelling environment provides a framework for building arbitrary process chains. The modeller can flexibly place process steps on the canvas via drag & drop from the OHLF-specific process-modelling library and AnyLogic's built-in library. AnyLogic's built-in library covers generic process-modelling objects such as delay and queue. The OHLF-specific process-modelling library extends AnyLogic's library with detailed models of lightweight manufacturing processes. Current machine models include an infrared oven, a hydraulic press, an injection moulding machine, resin transfer moulding injection unit and a pultrusion machine. Further manufacturing processes are planned to be modelled and integrated into the OHLF-specific process-modelling library.

The lower half of Fig. 5.6 illustrates the variation of the product's energy intensity over the production volume. The product energy intensity is broken down into direct and indirect energy demand. While the direct energy demand remains constant, the indirect energy demand and consequently the product energy intensity decrease with higher production volumes. This is due to the energy-intensive ramp-up phases at the beginning of production, which account for a smaller share of the part-specific energy demand at higher lot sizes. The upper part of Fig. 5.6 depicts the contribution of the three machines on the tunnel's energy intensity at different production volumes. Comparing the machine's shares at a lot size of ten and 75, it becomes apparent that the energy demand of the oven and injection moulding machine decreases while the energy demand of the press increases. This is due to the cycle time of the press, which is the smallest in the process chain. As a result, the press spends longer times in the standby mode waiting for parts to be processed then the injection moulding machine. The infrared oven, being the bottleneck of the process, has no time shares in the standby state.

A challenge in the implementation of the current approach is the provision of process parameters for the machine models. Process parameters have commonly not been set in early design stages. Therefore, the modelling engineer has to choose reference values based on own experiences or literature values. The extension of the modelling approach to numerical process simulation to exactly calculate process parameters would provide

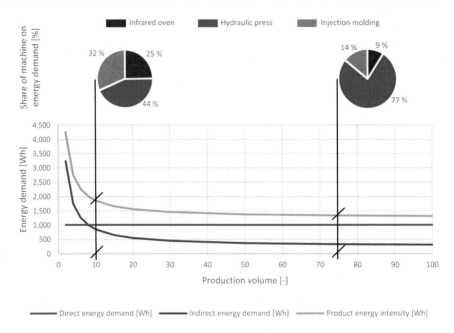

Fig. 5.6 Variation of energy demand and machine's share on product energy intensity depending on production volume

help in this regard. Further steps in developing the modelling approach could include the integration of more manufacturing processes to the OHLF-specific process-modelling library.

5.8   Summary and Outlook

The use of multi-material lightweight body parts in vehicles increases the environmental relevance of automotive component manufacturing. Assessing the energy intensity of the manufacturing of lightweight components in early design stages would enhance the possibilities of life cycle engineering. This chapter has illustrated a simulation modelling approach for new manufacturing processes and assessing their environmental impact with a focus on electrical energy demand. The modelling approach was prototypically implemented and its functionality demonstrated at the Open Hybrid LabFactory. A challenge for the application of the approach is the parameterization of the machine models. In early design stages, exact process and machine parameters are scarce. Therefore, reference values of similar products or estimations have to be used. A possible research direction for future work could be first the combination with numerical process simulation that provides process parameters for the machine model and secondly, extending the existing OHLF-specific process models with further manufacturing processes.

References

Abele E., Braun S., Schraml P. (2015). Holistic Simulation Environment for Energy Consumption Prediction of Machine Tools. Procedia CIRP 29:251–256. doi: https://doi.org/10.1016/j.procir.2015.02.059

Buschhoff C., Brecher C., Emonts M. (2016). High volume production of lightweight automotive structures. In: Bargende M., Reuss H.-C., Wiedemann J. (eds) 16. Internationales Stuttgarter Symposium. Springer Fachmedien Wiesbaden, Wiesbaden, pp. 213–226 https://doi.org/10.1007/978-3-658-13255-2_17

Der, Antal, et al. „Modelling and analysis of the energy intensity in polyacrylonitrile (PAN) precursor and carbon fibre manufacturing." Journal of cleaner producyion 303(2021):127105. https://doi.org/10.1016/j.jclepro.2021.120105

Dröder K., Herrmann C., Raatz A., Große T., Schönemann M., Löchte C. (2014) Symbiosis of plastics and metals: integrated manufacturing of functional lightweight structures in high-volume production. Kunststoffe im Automobilbau. Mannheim pp. 31–44 https://doi.org/10.1016/j.procir.2015.12.063

Duflou J.R., Deng Y., van Acker K., Dewulf W. (2012) Do fiber-reinforced polymer composites provide environmentally benign alternatives? A life-cycle-assessment-based study. MRS Bull. 37:374–382. doi: https://doi.org/10.1557/mrs.2012.33

Fleischer J., Teti R., Lanza G., Mativenga P., Möhring H.-C., Caggiano A. (2018) Composite materials parts manufacturing. CIRP Annals 67:603–626. doi: https://doi.org/10.1016/j.cirp.2018.05.005

Garwood T.L., Hughes B.R., Oates M.R., O'Connor D., Hughes R. (2018) A review of energy simulation tools for the manufacturing sector. Renewable and Sustainable Energy Reviews 81:895–911. doi: https://doi.org/10.1016/j.rser.2017.08.063

Herrmann C., Thiede S., Kara S., Hesselbach J. (2011) Energy oriented simulation of manufacturing systems – Concept and application. CIRP Annals 60:45–48. doi: https://doi.org/10.1016/j.cirp.2011.03.127

Herrmann C., Dewulf W., Hauschild M., Kaluza A., Kara S., Skerlos S. (2018) Life cycle engineering of lightweight structures. CIRP Annals 67:651–672. doi: https://doi.org/10.1016/j.cirp.2018.05.008

Ingarao G., Di Lorenzo R., Micari F. (2011) Sustainability issues in sheet metal forming processes: an overview. Journal of Cleaner Production 19:337–347. doi: https://doi.org/10.1016/j.jclepro.2010.10.005

Kaluza, A., Kleemann, S., Fröhlich, T., Herrmann, C., & Vietor, T. (2017). Concurrent Design & Life Cycle Engineering in Automotive Lightweight Component Development. Procedia CIRP, 66, 16–21. doi:https://doi.org/10.1016/j.procir.2017.03.293

Kaluza, Alexander; Hagen, Johanna Sophie; Der, Antal; Cerdas, Felipe; Herrmann, ChristophLife Cycle Assessment of Thermoplastic and Thermosetting CompositesIn: Rangappa, Sanjay Mavinkere; Parameswaranpillai, Jyotishkumar; Siengchin, Suchart; Kroll, Lothar, Lightweight Polymer Composite Structures. Design and Manufacturing Techniques., CRC Press/Taylor & Francis, Boca Raton, 2020, Seite 359-385, DOI 10.1201/9780429244087-13

Kellens K., Dewulf W., Overcash M., Hauschild M.Z., Duflou J.R. (2012) Methodology for systematic analysis and improvement of manufacturing unit process life-cycle inventory (UPLCI)—CO2PE! initiative (cooperative effort on process emissions in manufacturing). Part 1: Methodology description. Int J Life Cycle Assess 17:69–78. doi: https://doi.org/10.1007/s11367-011-0340-4

Li W. (2015) Efficiency of Manufacturing Processes. Sustainable Production, Life Cycle Engineering and Management. Cham: Springer International Publishing. https://doi.org/10.1007/978-3-319-40277-2.

Posselt G., Fischer J., Heinemann T., Thiede S., Alvandi S., Weinert N., Kara S., Herrmann C. (2014) Extending Energy Value Stream Models by the TBS Dimension – Applied on a Multi Product Process Chain in the Railway Industry. Procedia CIRP 15:80–85. doi: https://doi.org/10.1016/j.procir.2014.06.067

Schönemann M., Schmidt C., Herrmann C., Thiede S. (2016) Multi-level Modeling and Simulation of Manufacturing Systems for Lightweight Automotive Components. Procedia CIRP 41:1049–1054. doi: https://doi.org/10.1016/j.procir.2015.12.063

Schönemann M., Bockholt H., Thiede S., Kwade A., Herrmann C. (2019) Multiscale simulation approach for production systems. Int J Adv Manuf Technol 38:81. doi: https://doi.org/10.1007/s00170-018-3054-y

Seow Y., Rahimifard S. (2011) A framework for modelling energy consumption within manufacturing systems. CIRP Journal of Manufacturing Science and Technology 4:258–264. doi: https://doi.org/10.1016/j.cirpj.2011.03.007

Thiede S. (2012). Energy Efficiency in Manufacturing Systems. Sustainable Production, Life Cycle Engineering and Management. Cham: Springer International Publishing. https://doi.org/10.1007/978-3-319-40277-2.

Consideration of Environmental Impacts of Automotive Lightweight Body Parts During the Conceptual Design Stage

6

Alexander Kaluza, Andreas Genest, Tobias Steinert, Andreas Schiffleitner and Christoph Herrmann

Abstract

Designing lightweight hybrid multi-material parts for automotive body applications is subject to a large solution space. This results in a large variety of concept alternatives that are able to fulfil the technical performance criteria. Towards prioritizing eco-efficient concepts, life cycle environmental impacts need to be determined at an early design stage. Life cycle assessment (LCA) is a system analysis methodology that enables a quantitative analysis of energy and resource flows and associated environmental impacts. LCA models need to reflect the application and combination of different materials, manufacturing process chains and end-of-life treatments as well as different scenarios within the vehicle use stage. At the same time, LCA models need

A. Kaluza (✉) · C. Herrmann
Institute of Machine Tools and Production Technology (IWF), Technische Universität
Braunschweig, Braunschweig, Deutschland
e-mail: a.kaluza@tu-braunschweig.de

C. Herrmann
e-mail: c.herrmann@tu-braunschweig.de

A. Genest · T. Steinert
Ifu Hamburg GmbH, Hamburg, Deutschland
e-mail: a.genest@ifu.com

T. Steinert
e-mail: t.steinert@ifu.com

A. Schiffleitner
iPoint-Austria gmbh, Wien, Österreich
e-mail: andreas.schiffleitner@ipoint-austria.at

T. Vietor (ed.), *Life Cycle Design & Engineering of Lightweight Multi-Material Automotive Body Parts*, Zukunftstechnologien für den multifunktionalen Leichtbau,
https://doi.org/10.1007/978-3-662-65273-2_6

to cope with limited information availability at this early stage of design. Therefore, the current chapter provides an overview on methodological aspects in evaluating lightweight body part concepts for different life cycle scenarios. A software toolchain is implemented and applied in the course of two case studies.

6.1 Environmental Assessment of Lightweight Body Part Concepts

The MultiMaK2 project aimed at guiding design processes for automotive body parts at early stages of conceptual design towards identifying and prioritizing eco-efficient alternatives. The underlying design process brought forward redesigned body part concepts through introducing and combining different materials (steel, aluminium, fibre-reinforced plastics) in hybrid multi-material designs (Chap. 2). The development goal was to reduce body part mass while maintaining or improving mechanical performance (strength, stiffness, minimal deflection in case of crash). Considering the entire life cycle of a body part (raw materials provision, manufacturing, use stage and end-of-life), eco-efficient concepts should cause lower environmental impacts compared to a functional equivalent reference concept. Therefore, all concepts need to undergo a constant evaluation regarding their associated environmental impacts to identify hotspots and major trade-offs. The application of lightweight materials and semi-finished products is likely to cause increased environmental impacts in raw materials provision, manufacturing and end-of-life (Herrmann et al. 2018). However, mass reductions enable to reduce energy demands of vehicles during their use stage (Herrmann et al. 2018). Environmental impacts associated to those decreased energy demands differ between combustion engine and electrified vehicles. In the latter case, the electricity mix applied for vehicle charging has a major influence on the benefits of lightweight body parts (Egede 2017). Break-even analyses enable to identify the occurrence, quantity and relation of the described effects.

An LCA-based approach was developed and implemented to address the aforementioned challenges. Its objective is to enable a quick and robust comparison between different lightweight concept alternatives while considering multiple specifications for the different life cycle stages. As depicted in Fig. 6.1, this encompasses a distinctive modelling of material and energy flows when determining the life cycle inventory (LCI) as well as a consecutive life cycle impact assessment (LCIA) based on this this inventory.[1]

The foreground systems include energy and material flows directly related to the body part concepts. Thereby, different fore- and background systems of the techno-sphere have to be considered (Herrmann et al. 2018). Whereas raw materials, body part

---

[1] An extensive review on aspects and methods to evaluate environmental impacts of lightweight structures has been provided in (Herrmann et al. 2018). The present approach is an excerpt with a focus on conceptual design.

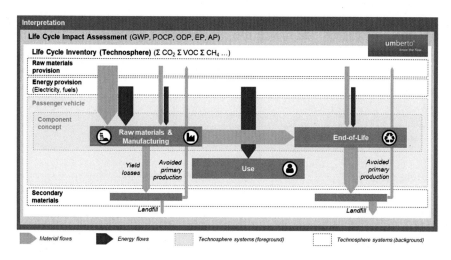

Fig. 6.1 Schematic illustration of material and energy flows considered within the semi-streamlined LCA approach

masses, manufacturing and end-of-life stages are directly linked to the body part itself, energy savings within the use stage need to be evaluated in the context of the surrounding passenger vehicle. Within the current study, both are part of the foreground system, as characteristics are directly influenced by body part design and mass. Efforts for providing raw materials and energy to manufacture the body part as well as to treat it in its end-of-life stage are part of the background system that cannot be directly influenced by design activities and decisions. Manufacturing and end-of-life modelling include material flows of secondary materials in an open-loop process. Electricity and fuel models follow a well-to-wheel (WTW) perspective that includes energy provision up to the point of charging/fuelling as well as drivetrain efficiency losses and driving emissions.

Due to unknown details of the product life cycle and shortcomings in availability of primary data in conceptual design, simplifications and generalizations in inventory modelling are required. The maturity of concepts could vary from early analytical models to numerical validated body parts (Chap. 2). The minimum requirements to evaluate environmental impacts include a bill of materials including mass, semi-finished products and major manufacturing processes. The model aims at serving as an early iteration LCA with the possibility to further detail sub-models with increasing knowledge of the product concept and its application.

In terms of life cycle impact assessment and interpretation, midpoint indicators were chosen to facilitate interpretability and comparability to other automotive LCA studies. Due to its prominent role in the scientific and public debate on global challenges, impacts on global warming potential (GWP) were evaluated upfront. In addition, an evaluation of further impact categories can be performed. In automotive LCA studies, most common midpoint categories comprise photochemical ozone creation potential (POCP),

ozone depletion potential (ODP), eutrophication potential (EP) and acidification potential (AP) (Broch 2017).

The following sections further detail the methodology and its general assumptions
based on the state of research and describe the model implementation. Case studies are
performed for a centre tunnel and a roof reinforcement structure.

6.2 Modelling of Systems within the Technosphere

Throughout the project, a library of modular and interconnected technosphere systems
has been modelled applying the software Umberto LCA +.[2] Methodological aspects are
discussed in the current section, whereas software implementation is in the focus of Sect.
6.3. Table 6.1 provides an overview on considered sub-systems. Those are aligned to
the life cycle stages of raw materials and manufacturing, use and end-of-life. As well,
energy provision is modelled with respect to fuel or electricity supplies to the vehicle
use stage. A combined evaluation of the sub-systems enables to describe and compare
different body part concepts throughout their life cycles. The GaBi SP30[3] and Ecoinvent
3.5[4] databases serve as an external source for background inventory data. Additional
inventory data was introduced based on the state of research as well as selected primary
data. In general, the modelling assumes manufacturing and end-of-life to take place in
Germany. The model depicts state-of-the-art technologies and assumptions for all life
cycle stages. Probable future developments influencing environmental impacts of lightweight structures are briefly discussed within the description of the single sub-systems.
However, a structured quantitative analysis of those developments will be a major objective for further research.

6.2.1 Raw Materials, Manufacturing and End-of-Life

The modelling of the initial life cycle stages incorporates a cradle-to-gate perspective
(see Chap. 5). A large design space can be observed for raw materials and semi-finished
products to manufacture hybrid multi-material body parts leading to a large variety of
materials, semi-finished products and manufacturing steps. In the course of the project,
this range was narrowed down to fundamentally suitable materials and manufacturing routes for shell designs in large series automotive applications. General trends for
steel alloys include the application of advanced high strength (AHSS) and ultra-high-
strength steels (UHSS). Aluminium profiles and sheet materials have found their way to

---

[2] https://www.ifu.com/umberto/oekobilanz-software/

[3] http://www.gabi-software.com/support/gabi/

[4] https://www.ecoinvent.org/database/ecoinvent-35/ecoinvent-35.html.

Table 6.1 Considered sub-systems within the technosphere as a foundation for environmental evaluation, based on (Kaluza et al. 2017)

Energy provision	
Fuels • Gasoline • Diesel	**Electricity** • Current market electricity mixes • Future scenarios based on REN21 targets

Raw materials & Manufacturing	Use	End-of-Life
Semi-finished products • Steel sheet (HSS, AHSS) • Aluminium sheet • GFRP & CFRP tapes & organosheets (PA6 matrix) • PA6 granulate, glass filled **Manufacturing** • Deep drawing • Stamping & bending • Thermoforming • Injection moulding • Yield losses	**Drivetrains** • Gasoline turbocharged • Diesel turbocharged • Battery electric **Drive Cycles** • WLTC • NEDC • Adapted Cycles **Lifetime driving distances** • Specific for each vehicle class	• Process route • EoL vehicle treatment • Shredding • Sorting • Application of recyclates • Avoided primary production • Landfill

medium and large-scale series production of higher-priced vehicles. At the same time, market demands for glass- or carbon fibre-reinforced plastics (GFRP, CFRP) are steadily increasing (Witten et al. 2018). Within the project, the focus was set on thermoplastics-based FRP. This includes organosheets as a replacement of steel and aluminium sheet materials as well as unidirectional (UD) tapes often applied both as building blocks of larger FRP structures and as pinpointed reinforcement to body parts to match needs of specific load cases. In addition, injection moulding enables to manufacture local reinforcement structures, e.g. cross ribs.

Steel sheet materials are assumed to originate from a blast furnace process. A value-corrected approach for representing different alloying elements and sheet thicknesses was applied based on the method presented by Broch (Broch 2017). Therefore, LCI data for different alloying elements was obtained based on GaBi SP30 as well as the works of Nuss et al. (Nuss et al. 2014). A mass-based inventory (alloy mixer) has been created, whereas lower sheet thicknesses were obtained based on additionally required rolling steps to be required to realize those thicknesses. With respect to alloy contents, chromium and nickel contents show the highest positive sensitivities to greenhouse gas emissions. Thus, an increased share in alloying leads to higher environmental impacts per kg material. Aluminium sheet materials were modelled as originating from primary manufacturing routes and again considering a mass-corrected approach for different alloying elements and sheet thicknesses. In the case of aluminium sheet production, highest sensitivities were found for the share of magnesium as an alloying element (see also (Broch 2017)). FRP semi-finished products have been modelled applying polyamide 6 and polyamide 6.6 as thermoplastic matrix materials. Fibre-mass contents served as a foundation to calculate the share between fibre reinforcement and matrix materials. Specific values for fibre-mass contents have been determined from material analyses of GFRP and CFRP UD tapes and organosheets in the course of the MultiMaK2 project. As information on manufacturing efforts for tapes and organosheets was not available within secondary databases, indications for electricity demands and material flows have been analysed from the state of research (Delogu et al. 2017; Hohmann and Schwab 2015; Hohmann et al. 2018). All sources identify that fibre and matrix material manufacturing represent the largest share of impacts from manufacturing of UD tapes and organosheets. However, heating processes seem to dominate the energy demands of process chains (Hohmann et al. 2018). Due to limited information availability and non-quality assured primary data from the Open Hybrid LabFactory at this stage, FRP manufacturing efforts are approximated for the current case using the physical models for pre-heating (Chap. 5). Further manufacturing processes, such as injection moulding, are described through secondary data.

In accordance to the findings of Broch and Hohmann et al. (Broch 2017; Hohmann et al. 2018), environmental impacts of process chains for both metallic and FRP materials are very sensitive to yield losses in part manufacturing. Therefore, scenarios describing several manufacturing yields were in this regard considered in this regard. For sheet materials, a standard scenario of 62.5% yield is estimated. Minimum and maximum scenarios are estimated with 50 and 75%, respectively (Broch 2017). The UD-tape process chain is assumed to achieve a yield of 90% due to its high adaptability to different geometries. As a result of expert interviews at Open Hybrid LabFactory, injection moulding is assumed with a 95% manufacturing yield.

The modelling of production waste and end-of-life treatment enables to determine if environmental impacts could be mitigated by processing wastes to secondary materials and in turn avoiding primary materials production. Thereby, closed- and open-loop processes are distinguished, whereas the first leads to recyclates being reintroduced to automotive manufacturing, the latter result in materials made available to other industries. In both approaches, efforts for collecting wastes and reprocessing secondary products or materials need to be taken into account when assessing the overall environmental impacts of end-of-life processes (Geyer et al. 2016). Another major challenge in closed- and open-loop processes for automotive high-volume manufacturing is the evaluation of available secondary material flow quantities (Herrmann et al. 2018). For the underlying study, production wastes are modelled as open-loop processes following an economic allocation for scrap steels and secondary aluminium. For FRP materials, feasibility has been proven for recycling and reintroducing recyclates to automotive manufacturing. A recent example has been presented by Gorbach et al. (Gorbach et al. 2018). However, there is no implementation in large-scale automotive manufacturing at this point of time. Thus, potential environmental benefits cannot be quantified for the presented study and is thus neglected. The end-of-life assessment is simplified as well. Efforts for processing end-of-life vehicles are estimated based on GaBi SP30. Scrap outputs are considered as input to electric arc steel manufacturing or aluminium ingot manufacturing. For FRP, production and end-of-life wastes are treated as input to thermal recovery. In turn, only virgin materials are considered assumed as input for manufacturing.

6.2.2   Use

Use stage modelling considers energy consumption of vehicles and possible reductions through lightweighting. Use stage energy demands of road vehicles are determined based on the works required to move the vehicles within a specific application as well as the efficiency of the vehicle drivetrain to provide this work. Driving forces to overcome include drag, elevation, rolling and acceleration forces. In the case of standardized driving cycles, only rolling and acceleration force are influenced by changes in vehicle mass (Herrmann et al. 2018). Within the state of research, it has been elaborated that effects of small changes in vehicle mass on a vehicle's fuel or energy demand can be expressed as engineering constants, so-called fuel or energy reduction values (FRV or ERV, respectively) (Lewis et al. 2014; Koffler and Rohde-Brandenburger 2010; Kim et al. 2010). FRV and ERV express the effect of a mass change when following a specific route over a given distance. FRV and ERV could include direct effects without an adaption of the

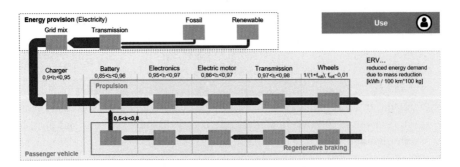

Fig. 6.2 Modelling of reduced environmental impacts due to weight reduction of battery electric vehicles based on (Liebl et al. 2014; Hofer 2014; thinkstep AG 2016) as well as parametrization from (Hawkins et al. 2012; Grunditz and Thiringer 2016; Campanari et al. 2009; Duce et al. 2013)

powertrain (primary effects) or including an adaption of the powertrain due to light-weighting (secondary effects).

The implemented model is based on the approach presented by Rohde-Brandenburger (Liebl et al. 2014) and has been extended using the works of Hofer and thinkstep (Hofer 2014; thinkstep AG 2016) as well as parameterizations from further publications (Hawkins et al. 2012; Grunditz and Thiringer 2016; Campanari et al. 2009; Duce et al. 2013). An overview on the developed model is presented in Fig. 6.2. Its focus is on primary mass reductions, as the developed body part concepts are estimated to achieve mass reductions of a maximum of 10 kg for each individual case compared to a reference design. Therefore, no powertrain adaptions are expected to occur as a direct effect from introducing one of the body parts. In MultiMaK2, the model has been refined to enable an evaluation of the standardized Worldwide Harmonized Test Cycle (WLTC) Class 3. Its major differences to the New European Driving Cycle (NEDC) lie in a more dynamic acceleration and deceleration behaviour as well as in increased top speeds, that intends to better represent real driving. Therefore, acceleration work to realize the vehicle movement is increasing in WLTC. At the same time, WLTC provides a larger share for an energy regeneration during deceleration stages. This occurs in cases where the vehicle's kinetic energy exceeds rolling and drag forces (Liebl et al. 2014). Further, the implemented model is prepared to be adapted to other driving cycles through parametrization. The second major design step to the model was made through adaption to represent battery electric vehicles. In contrast to combustion engines, regenerative braking has been introduced. This included an analysis of the WLTC deceleration sections that has been executed following the example of acceleration behaviour in the state of research (Hofer 2014; thinkstep AG 2016). In result, propulsion energy demands and energy recovery potentials could be quantified for the given drive cycle (energy balance at wheel).

The inventory modelling for both combustion and battery electric vehicles follows a well-to-wheel (WTW) approach. For combustion engines, this includes all upstream activities for providing gasoline and diesel fuels to a fuel station. Further, emissions from fuel combustion are determined. Whereas a direct link between greenhouse gas emissions and (avoided) fuel combustion can be drawn, the emission of other pollutants is as well majorly influenced by engine and exhaust treatment technologies (Herrmann et al. 2018; Pandian et al. 2009; Fiebig et al. 2014). Therefore, aside from greenhouse gas emissions, the implemented model relies on a mass-based allocation based on limit values from EU emission standards to determine shares of environmental impacts.

In the case of electric vehicles, the WTW perspective includes the generation of electricity, their transmission to the charging infrastructure, the charging itself as well as the energy conversion within the respective drivetrain. As described within the eLCAr guidelines for environmental assessment of electric vehicles, major body parts to include in drivetrain evaluation are the battery, power electronics, the electric motor as well the transmissions (Hofer 2014). Fig. 6.2 depicts the energy flows for propulsion as well as for regenerative braking. The calculations were performed using median values from reported drivetrain efficiencies of the single body parts as reported within the state of research (Hofer 2014; thinkstep AG 2016; Hawkins et al. 2012; Grunditz and Thiringer 2016; Campanari et al. 2009; Duce et al. 2013). As well, a factor k needs to be introduced into the model to reflect the drivetrain system behaviour in dealing with energy provided by regenerative braking. The above listed sources estimate that a share of 50 to 80% of the total regenerative energy could be used for vehicle charging as a result of battery management. However, it was found that assumptions for k lack empirical evidence for reviewed studies. Thus, the parameter variability is reported within the identified ERV. Table 6.2 summarizes FRV and ERV determined from the described models at the example of an average class A vehicle. Further, WTW climate change impacts were determined based on those FRV and ERV using background information from Umweltbundesamt (German

Table 6.2 Fuel and energy reduction values for gasoline, diesel and battery electric drivetrains evaluated for a class A vehicle (e.g. Volkswagen Golf VII/ eGolf); based on (Liebl et al. 2014; thinkstep AG 2016; Hofer et al. 2014), derivation of WTW climate change impacts based on Ecoinvent 3.5. DE = Germany, NO = Norway, AUS = Australia

Drivetrain/ Unit/ Electricity Mix			NEDC	WLTC
Combustion (Gasoline)	**l/100 km*100 kg**		0.15	0.17
	kg CO$_2$eq/100 km*100 kg (WTW)		0.44	0.49
Combustion (Diesel)	**l/100 km*100 kg**		0.12	0.14
	kg CO$_2$eq/100 km*100 kg (WTW)		0.35	0.40
Battery Electric	**kWh/100 km*100 kg**		0.55 − 0,62*	0.60 − 0.68*
	kg CO$_2$eq/100 km*100 kg (WTW)	**DE**	0.33 − 0.37*	0.36 − 0.41
		NO	~0.02	~0.02
		AUS	0.53 − 0.60	0.58 − 0.66

* depending on drivetrain efficiency

Environment Agency).[5] All drivetrains were evaluated for the German market. In general, greenhouse gas emissions resulting from a 100 kg mass decrease are lower for diesel-powered vehicles in comparison to gasoline vehicles. Depending on k, the benefit of mass reduction in electrified drivetrains is lower or higher than for diesel engines. Norway and Australia served as sample cases to show the potential variability of greenhouse gas emissions in specific markets. Due to a high share in renewable energy, the effect of mass reduction almost vanishes for the Norwegian situation. A contrary observation can be made for Australia due to a high share of fossil sources to generate electricity. Shares of energy carriers for different markets for the present time as well as past years can be determined from World Bank data. Country-based targets for increasing renewable electricity supply could be derived from other international data suppliers, e.g. REN21, an international policy organization in the field of renewable electricity supply that provides scenarios for potential developments (REN21 2016).

Lifetime driving distances have been calculated based on the recommendations by Weymar and Finkbeiner (Weymar and Finkbeiner 2016). As the reference vehicle for body part redesign is a class A compact car, 200.000 kms have served as a basis for break-even calculations. It needs to be noted that this scenario is recommended for average vehicles in Germany being used in established ownership and operating models. Adapted assumptions for new operating models, e.g. "Mobility-as-a-Service", need to be derived in future studies. Besides lifetime driving distances, especially real-world driving profiles are expected to change in future vehicle generations.

6.3   Implementation

The presented LCA methodology has been embedded in a Life Cycle Engineering workflow that combines different tools in conceptual design, LCA and visualization. Its goal is to enable an efficient workflow with quick iterations to determine possible environmental impacts of lightweight body part design alternatives for passenger vehicles. A special focus is set on hybrid component designs that are composed of steel, aluminium and FRP materials.

The toolchain is organized around a core LCI model that has been implemented using the software Umberto LCA+ (see Fig. 6.3). The surrounding tiers comprise LCIA as well as interpretation, e.g. in terms of comparing conceptual designs, through visualization. The LCI model follows a modular architecture consisting of different sub-models. Each sub-model describes energy and material flows of specific sections of lightweight body part life cycles and can be seamlessly combined with other sub-models to represent a full life cycle. Thus, the life cycle of a body part is composed of at least five stand-alone sub-models describing 1) cradle-to-gate processes up to the semi-finishes product, 2) the

---

[5] http://uba.co2-rechner.de/de_DE/

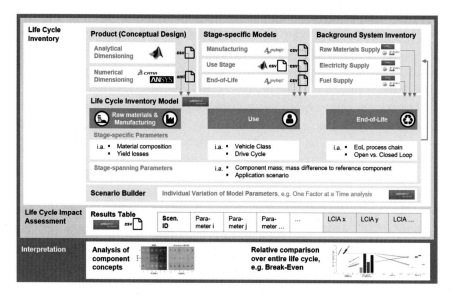

Fig. 6.3 Implemented toolchain for evaluating environmental impacts of lightweight components during conceptual design

gate-to-gate manufacturing, 3) use stage fuel or energy demands and 4) respective supply of fuel or electricity and 5) end-of-life treatment.

The implemented LCI modelling approach combines different advantages. Foremost, sub-models can be updated independent from each other. This enables to evolve the single aspects over a larger time horizon and future projects, e.g. through addition of further raw materials and semi-finished products, improved inventory data for innovative manufacturing processes or changing compositions of the electricity mix within the vehicle use stage. Further, the model can be used as a starting point for a full LCA study at later design stages. With increased knowledge of the product life cycle, all pre-set sub-models can be detailed and extended.

To evaluate a conceptual design, a life cycle model is composed from relevant sub-models and parameterized through part-specific information, e.g. mass of a material within a hybrid design. All parameterizations can be executed through standardized tabular inputs. This enables an efficient interface between the LCI model and models of other engineering disciplines. For example, conceptual design generates information on material compositions, masses and properties of semi-finished products. Structured exchange files (csv, xml) were created as a communication interface between the applied design & dimensioning tools (MATLAB, Catia V5 & ABAQUS) and the LCI model. The product-related parameterization is complemented with pre-set parameters for the stage-specific models, e.g. fuel reduction values (FRVs). As well, interfaces to simulation models developed for estimating yields and energy demands in manufacturing and end-of-life have been prepared. However, limited validation of the obtained model results hinders the deployment of this feature for robust LCA results at the current stage.

A scenario builder for the life cycle inventory model has been developed and extends the Umberto LCA+software package. It enables to define variations for parameters in the different inventory sub-models in a tabular interface. The scenario building can be automated by defining minimum, median and maximum values for specific inventory flows. With this approach, a larger number of conceptual designs can be parametrized and evaluated simultaneously with respect to different assumptions of the life cycle inventory model.

In terms of results visualization, different workflows have been realized. Impact assessment results for different scenarios and impact assessment methods are stored within a tabular database. Different tools that enable a visual exploration of impact assessment results can access this database. Break-even calculations have, for example, been implemented using the Python data visualization library Seaborn. Further approaches to enable an interaction with the obtained LCIA results are a research focus of the Life Cycle Design & Engineering Lab (see Chap. 7). The developed models have been applied within two case studies at different stages of conceptual design that will be presented in the following.

6.4   Case Study 1 – Roof Reinforcement

A conceptual design process of a roof reinforcement was executed based on the analytical dimensioning approach described in Chap. 2. The approach addresses a very early stage in body part development that brings forward initial design alternatives and evaluates their feasibility in terms of mechanical performance. Potential life cycle environmental impacts are determined and fed back to the development process to select most suitable design alternatives by considering specific life cycle scenarios. The executed design process is summarized in the upper part of Fig. 6.4. A U-shaped cross section geometry was analysed regarding its bending, tensile and torsional stiffness, strength and buckling stiffness. The design goal was to match performance of a reference design through variations of geometric parameters (width, height and wall thicknesses) as well as applied materials. While the analytical dimensioning approach would allow mixing materials, e.g. FRP and metallic materials, the current case study focuses on mono-material designs and their major performance differences at this early stage. The case study compares sheet materials from high-strength steels (HSS) and aluminium 6000 series as well as layered build-ups from unidirectional glass and carbon fibre-reinforced thermoplastic tapes (polyamide 6 reinforcement).

The lower part of Fig. 6.4 exemplarily shows results originating from analytical dimensioning while varying geometric parameters and applied materials. Mechanical performance was matched to a reference mild steel body part with a wall thickness of 1 mm and assuming constant flange widths as a requirement from joining. Design goals were set to match the strength of the reference body part by 90%, bending, elongation and torsional stiffness by 70% and buckling stiffness by 100%. The colour scale enables

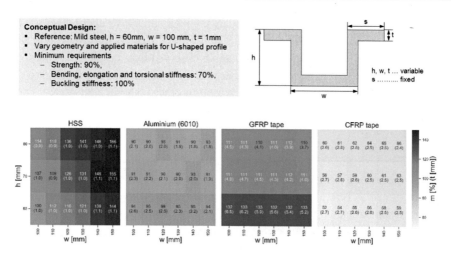

Fig. 6.4 Upper part – Approach for conceptual design of U-shaped profile by applying different materials; Lower part - Normalized minimum body part weights in relation to body part height (h), width (w) and sheet thickness (t)

to identify resulting masses with respect to widths and heights of the sample body parts for different materials. It has to be noticed that the applied design approach focuses on relative statements of body part concept masses. Thus, all values are normalized to a HSS design (width 100 mm, height 60 mm) that resembles alternatives in modern large-scale manufacturing. The parenthesized values additionally show necessary wall thicknesses to achieve performance criteria. It is apparent that CFRP designs overall show the lowest potential body part mass. GFRP designs could not achieve reference body part masses for the current case. However, increased profile heights could lower masses significantly. In the case of aluminium structures, decreased profile heights need to be compensated by increased wall thicknesses in order to match mechanical performance criteria. As a result, body part masses only show small variations for the observed design space. In the case of high-strength steel, a design option with the lowest resulting mass can be identified from the solution space.

Fig. 6.4 could further be used to identify the best alternative within an even more limited design space. For example, profile heights of roof reinforcement structures directly influence headroom on the vehicle's inside or frontal area at its outside. Limiting profile height to 60 mm would exclude of GFRP designs it the current case. However, for the current case study, a design space allowing widths up to 150 mm and heights up to 80 mm is considered. From that design space, lowest, median and highest mass options are determined for each of the material alternatives and represent a major input for life cycle environmental evaluation. The first step thereby includes the determination of potential environmental impacts from raw materials and manufacturing using the previously described model. The left part of Fig. 6.5 relates body part masses to greenhouse gas emissions from those life cycle stages. Steel and aluminium sheet materials are modelled to undergo a deep drawing process. GFRP and CFRP alternatives are built up from UD tapes and are processed by thermoforming. Manufacturing yields are estimated as median values (see Sect. 6.2). It can be observed that, despite showing lowest masses for the desired mechanical performance, CFRP concept alternatives show a comparatively high greenhouse impact in comparison to other material alternatives. For the current case, aluminium concepts are estimated to show a higher mass, with their greenhouse gas impact performing in the range of the CFRP. Steel and GFRP concepts lead to higher body part masses. Overall, lowest environmental impacts from raw materials and manufacturing are estimated to be achieved through HSS designs.

The body parts are assumed to be applied to a battery electric passenger vehicle. Use stage is assumed to resemble the WLTC test cycle over a typical mileage of 200.000 kms. Vehicles are assumed to be operated in Germany. Towards quantifying GHG emissions by applying a FRV, a reference mass and material needs to be assumed. In line with body parts that are currently applied in automotive high-volume production, the reference mass of a HSS part is estimated with 2 kg. The lower part of Fig. 6.5 compares the

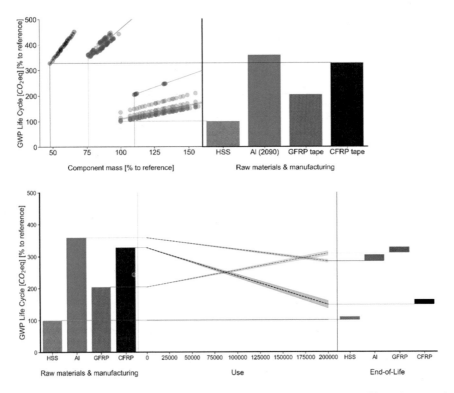

Fig. 6.5 Break-even calculation for electric vehicles GHG emissions over the life cycle, assuming a reference component with a mass of 2 kilograms HSS, Use stage: 200.000 kilometres, German electricity mix of 2018 with 0,6 kg CO2eq/kWh, WLTC, battery electric vehicle, end-of-life: only processing efforts

use stage of lowest mass alternatives for each material. As well, results variability due to varying energy reduction values is depicted, as roof reinforcement structures could be applied to vehicles with different drivetrain efficiencies. Deviating from the description in Sect. 6.2, end-of-life modelling only covers established collection and shredding efforts but does not account credits for reintroducing materials into the open-loop markets.

Overall, it can be observed that none of the aluminium, GFRP or CFRP alternatives can achieve a break-even during the assumed vehicle use. GFRP alternatives are heavier compared to the reference and cannot achieve a break-even. However, the CFRP design shows the highest mass reduction and thus largest use stage benefit in comparison to the reference design. Within a prolonged lifetime mileage, as it could be expected for Mobility-as-a-Service vehicles, a break-even is likely to be reached between 200.000 and 300.000 kms (Volkswagen AG 2018).

6.5   Case Study 2 – Centre Tunnel

The second case study evaluates life cycle environmental impacts of centre tunnel concepts. As observed for the roof reinforcement structure, the reference body part originates from the SuperLIGHT-CAR project. In contrast to case study 1, different semi-finished products are combined in single or multi-material designs. All concept alternatives fulfil basic feasibility criteria regarding cost and manufacturability in large series applications. The evaluation on mechanical performance depends on deformations observed in a crash load case. Only concepts that did not show any deformation within the numerical validation represent feasible alternatives. Towards that goal, the design and simulation engineering team (lead by IAV Automotive Engineering GmbH) performed an iterative improvement of all concepts resulting in eight conceptual designs. Those serve as an input for the implemented LCE toolchain. Table 6.3 lists all concept alternatives and potential mass reductions compared to the reference design (hot forming steel). Six concepts act as direct replacements of the reference body part. Concept 1 relies on thermoformed carbon fibre-reinforced plastics. Concepts 2 to 7 apply carbon fibre-reinforced UD tapes to support structures from press hardened steel, aluminium and glass-fibre-reinforced plastics. Injection-moulded glass-reinforced rib structures are applied for concepts 5 and 7. Mass reduction potentials range from 0.7 to 5 kg, depending on the specific alternative.

Life cycle environmental impacts for all conceptual designs are evaluated applying the previously described LCA model. Three different scenarios are estimated for manufacturing yields of all sheet materials (50%; 62.5%; 75%), whereas yields of tape materials and injection moulding processes are estimated to reach 95%. The handling of production scraps is modelled as an open-loop process. In line with the first case study, end-of-life covers scrap collection and shredding efforts. Use stage impacts are determined for combustion engine drivetrains only. Vehicle designs for fully electric

Table 6.3 Concept alternatives and potential mass reductions

No.	Type	Mass reduction [kg]	Materials						
			Hot forming steel	Press-hardening steel	Aluminium 6082	CFRP PA66 (organo sheet)	CFRP PA6-CF (UD patch)	GFRP PA6-GF (organo sheet)	GFRP PA6-GF30 (cross ribs)
R	Reference	0	■						
1		-5.0				■			
2		-1.8		■				■	
4		-2.7			■			■	
5		-3.0						■	
6		-0.7		■				■	
7		-1.0		■				■	

drivetrains typically do not include a centre tunnel, as its main function is to shield the exhaust system and the cardan shaft for the share of vehicles with rear or all-wheel drive.

Fig. 6.6 shows environmental impacts for all scenarios at the example of Global Warming Potential for the concepts that directly substitute the reference design. Out of six concepts, one aluminium-intensive and two CFRP-intensive concepts show significantly higher environmental impacts from raw materials and manufacturing stages and exceed the reference body part by 50 to 100%. The error bars thereby represent different scenarios with respect to manufacturing yields of the sheet semi-finished products. Resulting greenhouse gas emissions are higher for low production yields. Assuming a medium scenario for manufacturing yields, use stage emission reductions due to

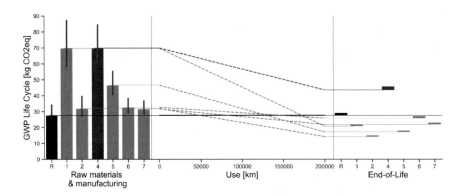

Fig. 6.6 Break-even analysis for centre tunnel lightweight concepts as direct substitution to a reference body part; GHG emissions over the life cycle, assuming a reference component with a mass of 7.5 kg HSS, use stage: 200.000 kms, Germany, WLTC, gasoline vehicle, end-of-life: only processing efforts

decreased body part mass are determined. Despite concept 4, all alternatives would reach a break-even point compared to the reference within a typical vehicle lifetime of 200.000 kms. While concepts 2, 6 and 7 show comparable efforts from raw materials provision and manufacturing, as their building blocks from steel and CFRP patches are similar to each other, concept 1 benefits from the highest mass reduction. Because of its modelling, the evaluation of the end-of-life stage does not affect the prioritization of concepts.

6.6   Summary and Opportunities

The present chapter addresses the evaluation of environmental impacts associated with automotive lightweight body part concepts. An LCA approach is followed that considers the entire body part life cycle. The inventory model is based on pre-defined sub-models that can be combined and parameterized to represent the assumed life cycle of body part concepts incorporating steel, aluminium and FRP materials. The implementation has been realized using the Umberto LCA + software. A newly introduced scenario builder helps to quickly evaluate the effects of parameter changes due to missing specifications in conceptual design. Software interfaces that facilitate the use of engineering data within the LCA model have been realized.

The methodology and implementation has been tested within two case studies that target the redesign of steel body parts with either single or multi-material approaches. Compared to the reference designs, all concept alternatives showed increased greenhouse gas emissions from raw materials provision and manufacturing stages. Due to tight constraints on design changes, the observed burden was significantly larger for mono-material concepts. The use stage of the roof reinforcement concepts has been evaluated as part of a battery electric vehicle being operated in Germany. In combination of raw materials, manufacturing and use stages, none of the concepts reaches a break-even for greenhouse gas emissions within the assumed vehicle use. This observation will be even more significant with increasing shares of renewable energy applied for vehicle charging. In contrast, centre tunnel concepts that incorporate multi-material designs could achieve a break-even compared to the reference design during the use stage of a gasoline engine vehicle. A major reason is a more effective and efficient material use through leveraging of material properties in line with mechanical loads. As the end-of-life model was limited to efforts for scrap handling, it did not influence the ranking.

The software implementation will be extended and refined as part of further research projects and case studies. There are several important streams for further research. First, the current sub-models for the different life cycle stages need to be refined. Manufacturing process models will benefit from increasing knowledge and process data of industrial scale processes. Changes in vehicle use will need to be reflected alongside the ongoing transformation of passenger vehicle operation. Prolonged lifetime mileages, adapted acceleration and deceleration behaviours or adapted vehicle concepts could influence the expected environmental impacts significantly. Further, repair and remanufacturing processes will gain importance as a measure to extend body part life cycles. As well, concepts integrating additional functions such as acoustic dampening or thermal insulation could reduce material demands. However, competing technical performance properties need to be accounted for.

References

Broch, F. (2017). Integration von ökologischen Lebenswegbewertungen in Fahrzeugentwicklungsprozesse. Wolfsburg: Springer. doi:https://doi.org/10.1007/978-3-658-18218-2

Campanari, S., Manzolini, G., & Garcia de la Iglesia, F. (2009). Energy analysis of electric vehicles using batteries or fuel cells through well-to-wheel driving cycle simulations. Journal of Power Sources, 186(2), 464–477. doi:https://doi.org/10.1016/j.jpowsour.2008.09.115

Delogu, M., Zanchi, L., Dattilo, C. A. A., & Pierini, M. (2017). Innovative composites and hybrid materials for electric vehicles lightweight design in a sustainability perspective. Materials Today Communications, 13(July 2018), 192–209. doi:https://doi.org/10.1016/j. mtcomm.2017.09.012

Duce, A. Del, Egede, P., Öhlschläger, G., Dettmer, T., Althaus, H.-J., Bütler, T., … Szczechowicz, E. (2013). Guidelines for the LCA of electric vehicles. Retrieved from http://www.elcar-project. eu/fileadmin/dokumente/Guideline_versions/eLCAr_guidelines.pdf (last visited 17.12.2019)

Egede, P. (2017). Environmental Assessment of Lightweight Electric Vehicles. Sustainable Production, Life Cycle Engineering and Management. Cham: Springer International Publishing. https://doi.org/10.1007/978-3-319-40277-2.

Fiebig, M., Wiartalla, A., Holderbaum, B., & Kiesow, S. (2014). Particulate emissions from diesel engines: correlation between engine technology and emissions. Journal of Occupational Medicine and Toxicology, 9(1), 6. doi:https://doi.org/10.1186/1745-6673-9-6

Geyer, R., Kuczenski, B., Zink, T., & Henderson, A. (2016). Common Misconceptions about Recycling. Journal of Industrial Ecology, 20(5), 1010–1017. doi:https://doi.org/10.1111/jiec.12355

Gorbach, G., Daberger, C., Föhner, A.-C., Gude, M., Luft, J., Troschitz, J., … Hüning, A.-C. (2018). ReLei Abschlussbericht. Retrieved from https://plattform-forel.de/relei-bericht/

Grunditz, E. A., & Thiringer, T. (2016). Performance Analysis of Current BEVs Based on a Comprehensive Review of Specifications. IEEE Transactions on Transportation Electrification, 2(3), 270–289. doi:https://doi.org/10.1109/TTE.2016.2571783

Hawkins, T. R., Gausen, O. M., & Strømman, A. H. (2012). Environmental impacts of hybrid and electric vehicles—a review. The International Journal of Life Cycle Assessment, 17(8), 997–1014. doi:https://doi.org/10.1007/s11367-012-0440-9

Herrmann, C., Dewulf, W., Hauschild, M., Kaluza, A., Kara, S., Skerlos, S., & Engineering, L. C. (2018). Life cycle engineering of lightweight structures. CIRP Annals, 67(2), 651–672. doi:https://doi.org/10.1016/j.cirp.2018.05.008

Hofer, J. (2014). Sustainability Assessment of Passenger Vehicles: Analysis of Past Trends and Future Impacts of Electric Powertrains. ETH Zurich, doi:https://doi.org/10.3929/ethz-a-010252775

Hofer, J., Wilhelm, E., & Schenler, W. (2014). Comparing the Mass, Energy, and Cost Effects of Lightweighting in Conventional and Electric Passenger Vehicles. Journal of Sustainable Development of Energy, Water and Environment Systems, 2(3), 284–295. doi:https://doi. org/10.13044/j.sdewes.2014.02.0023

Hohmann, A., & Schwab, B. (2015). MAI ENVIRO. Vorstudie zur Lebenszyklusanalyse mit ökobilanzieller Bewertung relevanter Fertigungsprozessketten für CFK-Strukturen. ISBN: 978–3–8396–0929–3

Hohmann, A., Albrecht, S., Lindner, J. P., Wehner, D., Kugler, M., Prenzel, T., … Reden, T. Von. (2018). Recommendations for resource efficient and environmentally responsible manufacturing of CFRP products Results of the Research Study MAI Enviro 2.0. ISBN: 978–3–9818900–0–6

Kaluza, A., Kleemann, S., Fröhlich, T., Herrmann, C., & Vietor, T. (2017). Concurrent Design & Life Cycle Engineering in Automotive Lightweight Component Development. Procedia CIRP, 66, 16–21. doi:https://doi.org/10.1016/j.procir.2017.03.293

Kim, H.-J., McMillan, C., Keoleian, G. A., & Skerlos, S. J. (2010). Greenhouse gas emissions payback for lightweighted vehicles using aluminum and high-strength steel. Journal of Industrial Ecology, 14(6), 929–946. doi:https://doi.org/10.1111/j.1530-9290.2010.00283.x

Koffler, C., & Rohde-Brandenburger, K. (2010). On the calculation of fuel savings through lightweight design in automotive life cycle assessments. International Journal of Life Cycle Assessment, 15(1), 128–135. doi:https://doi.org/10.1007/s11367-009-0127-z

Lewis, A. M., Kelly, J. C., & Keoleian, G. A. (2014). Vehicle lightweighting vs. electrification: Life cycle energy and GHG emissions results for diverse powertrain vehicles. Applied Energy, 126, 13–20. doi:https://doi.org/10.1016/j.apenergy.2014.03.023

Liebl, J., Lederer, M., Rohde-Brandenburger, K., Biermann, J.-W., Roth, M., Schäfer, H., … Schäfer, H. (2014). Energiemanagement im Kraftfahrzeug. doi:https://doi.org/10.1007/978-3-658-04451-0

Nuss, P., & Eckelman, M. J. (2014). Life cycle assessment of metals: A scientific synthesis. PLoS ONE, 9(7), 1–12. doi:https://doi.org/10.1371/journal.pone.0101298

Pandian, S., Gokhale, S., & Ghoshal, A. K. (2009). Evaluating effects of traffic and vehicle characteristics on vehicular emissions near traffic intersections. Transportation Research Part D: Transport and Environment, 14(3), 180–196. doi:https://doi.org/10.1016/j.trd.2008.12.001

REN21. (2016). Renewables 2016 Global Status Report. Paris. Retrieved from https://www.ren21. net › 2019/05 › REN21_GSR2016_FullReport_en_11 (last visited 17.12.2019)

thinkstep AG. (2016). Report on LCA results for utilization phase model. Retrieved from http://www.project-alive.eu/pdf/d6-5-report-on-lca-results-for-utilization-phase-model.pdf (last visited 17.12.2019)

Volkswagen AG. (2018). Leadership in Mobility-as-a-Service (MaaS). Retrieved from https://www.volkswagenag.com/presence/investorrelation/publications/presentations/2018/01_january/2018-01-10_Presentation_Johann_Jungwirth_Las Vegas.pdf (last visited 17.12.2019)

Weymar, E., & Finkbeiner, M. (2016). Statistical analysis of empirical lifetime mileage data for automotive LCA. International Journal of Life Cycle Assessment, 21(2), 215–223. doi:https://doi.org/10.1007/s11367-015-1020-6

Witten, E., Mathes, V., Sauer, M., & Kühnel, M. (2018). Composites Market Report 2018 : Market developments, trends, outlooks and challenges, (September), 1–44. Retrieved from https://www.avk-tv.de/files/20181115_avk_ccev_market_report_2018_final.pdf (last visited 17.12.2019)

Life Cycle Design and Engineering Lab in the Open Hybrid LabFactory

7

Alexander Kaluza, Sebastian Gellrich, Sebastian Thiede and Christoph Herrmann

Abstract

Engineering processes for innovative and eco-efficient automotive components show a high degree of labour division. Domain-specific information needs to be exchanged between actors and serves as input for decision-making, e.g. information on part performance, weight, cost or environmental impact. In current engineering practice, this cross-domain communication tends to be streamlined up to the level of selected and simplified KPI that represent the progress of individual disciplines. This hinders a holistic improvement of products and processes. Research within the MultiMaK2 project emphasizes the importance of a joint knowledge building between engineering disciplines and aims at creating a cross-domain understanding of root causes for hotspots and goal conflicts. Therefore, the Life Cycle Design & Engineering Lab was established at the Open Hybrid LabFactory. It objectifies the methodological approach of visual analytics through domain spanning software toolchains, centralized data acquisition, analytics methods as well as a variety of visualization tools and hardware elements that serve the described goals.

A. Kaluza (✉) · S. Gellrich · S. Thiede · C. Herrmann
Institute of Machine Tools and Production Technology (IWF), Braunschweig, Deutschland
e-mail: a.kaluza@tu-braunschweig.de

S. Gellrich
e-mail: s.gellrich@tu-braunschweig.de

S. Thiede
e-mail: s.thiede@tu-braunschweig.de
C. Herrmann
e-mail: c.herrmann@tu-braunschweig.de

© Springer-Verlag GmbH Germany, part of Springer Nature 2023
T. Vietor (ed.), *Life Cycle Design & Engineering of Lightweight Multi-Material Automotive Body Parts*, Zukunftstechnologien für den multifunktionalen Leichtbau,
https://doi.org/10.1007/978-3-662-65273-2_7

7.1   Automotive Life Cycle Engineering from the Open Hybrid Lab Factory's (OHLF) Perspective

The mitigation of negative environmental impacts of their products and processes is a major concern for the automotive industry. This encompasses the entire life cycle of vehicles, including raw materials, manufacturing, use and end-of-life stages. One example is the Volkswagen AG that announced to achieve a CO2-neutral mobility up to 2050 with an intermediate goal of reducing greenhouse gas emissions by 30% between 2015 and 2025 (Volkswagen AG n.d.). Against that background, new vehicle technologies need to be engineered, incorporating reduction targets for environmental impacts. Life cycle engineering (LCE) is a means to guide engineering processes with respect to overarching sustainability goals (Hauschild et al. 2017). Life Cycle Design & Engineering (LCDE) refers to a close link between engineering activities and their impact on the entire life cycle. At Open Hybrid LabFactory (OHLF), LCDE support has two starting points. First, engineering processes can be targeted that are in the direct focus of engineers at OHLF (foreground system). This includes the design and manufacturing of innovative body parts (see Fig. 7.1). Research in manufacturing could promote innovative designs, e.g. the combination of two materials on component level (technology push). In turn, adapted manufacturing processes could result from adapted design requirements (market pull). Second, design and manufacturing research at OHLF influences the entire vehicle life cycle and is, vice versa, influenced by the different life cycle stages. For instance, new component designs could pose challenges to recycling. In turn, requirements from those life cycle stages, e.g. expected vehicle lifetimes, can be translated to design requirements. LCDE therefore takes a cradle-to-grave life cycle perspective.

Within LCDE, life cycle assessment (LCA) serves as a foundational methodology to quantify environmental impacts of products and processes. However, the interface between the LCA method to product- or process-related engineering is cumbersome in practice. LCA requires expert knowledge to execute the method itself, including data

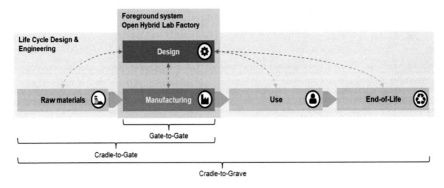

Fig. 7.1 System perspectives in Life Cycle Design & Engineering of automotive lightweight body parts at Open Hybrid LabFactory

acquisition and modelling as well as to interpret its results. This originates from complex interdependencies within the material and energy flows of a products' life cycle and multiple resulting impacts. At the same time, the scopes of engineering domains involved are rather. While domain-specific engineering decisions influence other life cycle stages, e.g. manufacturing cut-offs that affect waste streams, this cross-link is not emphasized in vehicle engineering.

Challenges for LCDE of vehicles increase with the shift to electric vehicles (EV) and new business models. For example, the effect of weight reduction on the use stage can be quantified with a low variability over different time horizons and geographic regions for ICEV. For EV, information on electric energy sources needs to be considered. However, electricity sources differ for every country and vary over time with increasing renewables in the supply (Egede 2017). Thus, if vehicle body parts are designed for different markets and one or more vehicle generations, no unambiguous statement on potential use stage benefits can be provided to the body part engineering teams. Therefore, decisions on favourable concepts are hampered.

The engineering of manufacturing technologies at OHLF ranges from laboratory to semi-industrial and industrial scales. In the sense of LCDE, this enables to design key process characteristics as well as to assess associated data, e.g. on preferable process windows, resulting cost, quality, time and associated environmental impacts. LCDE at OHLF enables to explore potential trade-offs and direct development, e.g. towards efficiency gains.

7.2   Background

Several research demands have been identified in relation to enhancing the application of LCA-based LCDE within previous research (Kaluza et al. 2018, 2019). This encompasses the identification of hotspots across different life cycle stages, impact categories, or sub-systems of a product, the comparison of two or more products or technologies, the identification of trade-offs, the assessment of technological, geographic or temporal variability as well as the identification of engineering levers to influence environmental and cost impacts. Table 7.1 presents a reworked summary of research demands based on previously published articles.

7.3   Understanding LCE through the Eyes of VA

When bringing together the challenges of LCDE with the goals of visual analytics (VA), potential synergies emerge. VA can be described as "the science of analytical reasoning facilitated by interactive visual interfaces" (Thomas and Cook 2005). VA is a human-centred process that enables the forming and testing of hypotheses and intends to reduce complex cognitive (engineering) work to process large data sets towards an

Table 7.1 Research demands for improving the application of LCA in engineering contexts, based on (Kaluza et al. 2018, 2019)

Challenge	Description
Strengthen prospective application of LCA	• LCA relies on quantified inventory data that is typically assessed for past time periods (retrospective character) • Additional methods and tools need to be introduced to allow prediction regarding specific life cycle stages or parameters, e.g. simulation- or data-based forecasting, scenario analysis
Increase accessibility of complex LCA models	• LCDE tools that streamlined LCA emerged in the past, e.g. linearized calculation methods, acceptance of data gaps, simplified and/or aggregated KPI (Rossi et al. 2016) • Inherent complexity of LCA results is neglected in simplified tools and hinders to identify complex mitigation strategies • Advanced computational LCA models enable to overcome this challenge, as e.g. shown by Cerdas et al. (2018)
Integrate different engineering scopes	• Certain levers to decrease environmental impacts are not in the engineering scope of the target audience, e.g. parallel engineering activities targeting • Other vehicle sub-systems and/ or • Upstream or downstream life cycle stages
Leverage primary (real-time) data	• Hard- and software for collecting primary product and process data is improving and can be leveraged for LCA • Acquisition, storage and analysis of life cycle related data • Processing and real-time feedback (Cerdas et al. 2017) • Goal: improvement and tailoring of inventory databases through generation of specific life cycle inventory data, e.g • Live mapping of energy and resource flows from manufacturing stage to the environmental impact (Cerdas et al. 2017) • Leveraging of use stage data for different user behaviours, e.g. driving patterns or geographic conditions (Li et al. 2016)
Provide tailored visualizations	• Current approaches for interpreting LCA studies tend to fail the demands of engineers and decision-makers (Laurin et al. 2016) • Potential reasons: overcomplexity, oversimplification, missing standards in results presentation, individual preferences, different target audiences, weak recommendation for action, e.g. feedback to adapt design parameters

informed decision-making (Kohlhammer et al. 2011). VA methods empower users to handle massive, dynamically changing data sets, detect expected and especially unexpected events, e.g. anomalies, changes, patterns and relationships, in order to gain new knowledge (Cook et al. 2007). Keim et al. structured constituting elements and processes of VA by describing the interplay of data acquisition, models, visualizations and knowledge building (Keim et al. 2009). The process has been adapted to the LCA methodology (Fig. 7.2) (Kaluza et al. 2018). In parallel to the key activities of VA, the analogies to an LCA-based LCE support are elaborated. These encompass inventory data acquisition, modelling, visualization and interpretation as well as the derivation of knowledge, as described in the previous section. A focus is set on challenges in performing and connecting the required activities with state-of-the-art methods and tools.

Data Inventory data builds the basis for any LCA study. Typically, studies combine primary and secondary data sources according to the goal and scope. Primary data can result from dedicated assessment campaigns or business information systems; secondary data sources mainly encompass commercially or publicly available inventory datasets and research studies. Pre-processing is a main task at the data stage. Primary data treatment requires activities like data cleansing, normalization, transformation as well as feature extraction. With respect to secondary inventory data, the challenge lies in the selection of appropriate datasets, e.g. with respect to key characteristics, system boundaries or spatial contexts.

Modelling LCA studies require a modelling of energy and resource flows, depletion of resources and emissions associated to a product or process of interest. LCA modelling relies on the integration of different domains and respective engineering models to map different life cycle stages. Dedicated software tools assist the inventory modelling. Overall inventory flows serve as an input for impact assessment that allows the derivation of environmental impacts, e.g. greenhouse gas emissions (GHG) measured in CO_2-eq. Other functionalities are the variation of models through sensitivity analyses

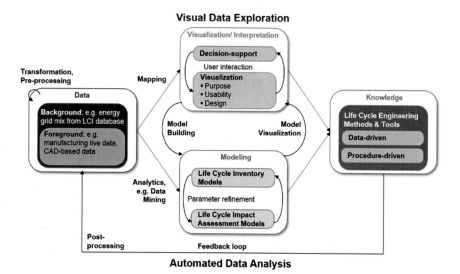

Fig. 7.2 Framework for understanding life cycle engineering through the eyes of visual analytics, previously published in (Kaluza et al. 2018), adapted from (Keim et al. 2008)

or the structured analyses of uncertainties. While traditional LCA tools enable a rather static modelling, dynamic system behaviours, e.g. in manufacturing, might be determined by specific engineering tools. This encompasses simulation-based as well as data-based methods.

Visualization/ Interpretation A major motivation of LCE is to translate LCA insights to engineering measures and decisions, ranging from ad hoc feedback within the engineering process up to decisions on a management or policy level. Life cycle impact assessment forms the basis for the interpretation of LCA results. In line with the listed insights of LCA studies as listed in Chap. 6, different visualizations can be chosen. Dedicated LCA tools provide visualizations that enable one or more of the described functionalities. However, on one hand this covers a high level of detail where high efforts are required to identify the relevant information for a given task. On the other hand, aggregated visualizations are incorporated at the level of non-experts. While allowing a quick interpretation, information on system dependencies is lost. Another stream is the representation of inventories, e.g. Sankey diagrams. In general, static visualizations dominate current LCA tools.

Knowledge A general distinction can be drawn between explicit and tacit knowledge derived from LCA studies. The management of explicit knowledge is very common in industrial and policy practice, e.g. by applying fixed rules. However, identifying and imparting tacit knowledge is a key challenge for every organization (Haldin-Herrgard 2000). LCA results typically allow case-dependent statements on the environmental impacts of product systems: "If product A is applied under the given circumstances, then the life cycle impact will be lower than for product B". This complexity leads to a translation of insights from LCA studies into domain- and application-specific methods and tools. The cumulated insights accelerate the LCE process for those specific domains. As well, continuous knowledge generation enables to enhance modelling and decision support.

7.4 The Life Cycle Design Engineering Lab (LCDEL)

The Life Cycle Design & Engineering Lab (LCDEL) has been initially set up with hard- and software to objectify the presented VA workflow and bundle research on life cycle-oriented automotive product and process engineering as well as digitalization research at OHLF. LCDEL is located at the shop floor level with a direct interface to manufacturing operations and analytics.

Three strategies are inherently linked to LCDEL's set-up and operation. First, enabling a life cycle perspective is seen as one of the key potentials as well as a necessity in engineering of future vehicle technologies. Research at OHLF focuses on gate-to-gate processes within the automotive life cycle, bringing forward innovative designs, materials and manufacturing processes. By providing insights from raw materials extraction, use stage and end-of-life, OHLF's engineering activities could be guided with those stages in mind. Second, a constant transfer of research findings to industrial practice is promoted through LCDEL. It provides state-of-the-art methods, hardware and software tools, ranging from commercialized solutions to scientific prototypes. LCDEL enables to initialize research activities in collaborative projects between academia and industry. The third strategy covers the exploration of engineering tools and technologies. Engineering research is simultaneously driven by technological innovations (pull) and brings forward innovative technologies at the same time (push). Therefore, a broad variety of hardware and software is provided and constantly updated.

Based on the presented strategies, three major application scenarios have been derived for LCDEL (see Table 7.2). Those cover the engineering of innovative automotive parts and their manufacturing technologies, the functionality as a Nerve Centre for OHLF's manufacturing engineering activities as well as a location assisting the progress of engineering meetings and review meetings on different decision levels.

The application scenarios emphasize LCDEL's character as a platform for performing research activities across scientific domains as well as to communicate research progress and key results between researchers and to decision-makers in industrial or policy contexts. The operation of LCDEL is strongly linked to the project portfolio of OHLF that covers short- and long-term projects solving current industrial demands (high TRL), collaborative industrial and academic research (medium TRL) as well as well as fundamental research activities (low TRL).

Table 7.2 Application scenarios of LCDEL at Open Hybrid LabFactory

Application scenario	Sub-elements
1. **Integrated engineering** of innovative automotive body parts and manufacturing technologies	• Conceptual design of innovative body parts • Planning of manufacturing process chains for multi-material components • Life cycle and cost assessment • Integration of engineering activities
2. **Nerve Centre** for OHLF manufacturing engineering	• Monitoring of live data • Process level: control parameters, process data, energy and material flows, part quality • Technical building services (abatement, climate control): control parameters, process data, energy demands • Process control and improvement based on live data (quality, time, cost, environmental impact) • Data- and simulation-based insights • Automated process control (CPPS)
3. **Engineering meetings and reviews**	• Creative working environment • Flexible configuration for different group sizes • Suitable for engineering activities as well as high level decision-making • Leveraging the potential of visualization

Table 7.3 Constituting elements of the Life Cycle Design & Engineering Lab (LCDEL)

VA	Elements	Implementation
D	Energy data acquisition and storage (OHLF)	- Real-time and historic energy data for individual manufacturing processes and technical building services - Siemens PAC, SENTRON Powermanager - OPC-DA Client/Server - WinCC 15.1 professional
	Process data acquisition and storage	- Real-time and historic process data of manufacturing processes, PLC access - Modbus, OPC UA - WinCC 15.1 professional
	Life cycle inventory (Secondary data)	- Energy & material flows for processes out of scope of OHLF - thinkstep GaBi SP38 (Professional + Extension DB) - Ecoinvent 3.6 - Further inventory data collection (state of research, primary)
M	Data & simulation-based modelling	- OHLF (gate-to-gate): - Goals: process understanding, prediction, efficiency improvements, scale-up behaviour, process planning, assessment of energy and resource flows - Outside OHLF (cradle-to-gate, gate-to-grave): - Goals: representation of upstream and downstream processes, e.g. raw materials sourcing, use scenarios, … - Data-based modeling: Python, Node RED, KNIME, … - Simulation-based modeling: Matlab Simulink, Anylogic, Plant Simulation,...
	Life cycle assessment	- Environmental impact assessment based on energy and material flows - Integration of sub-models from data- and simulation-based modelling - Tools: Umberto LCA+, Open LCA, thinkstep GaBi, Brightway2
V/I		- 4*4 Screen multi-input video wall, Barco Clickshare - Microsoft Surface Hub Interactive Display - Mixed Reality Head-Mounted device Microsoft HoloLens - Virtual Reality Head-Mounted devices, e.g. HTC Vive - Tools: Business Intelligence, Data Visualization, Interactive Visualization,...
	General lab equipment	- Flexible tables, chairs, high desks, variable partition walls to suit different workshop and review situations - Writable wall panels (6*2 meters)

The current hardware implementation of LCDEL is listed within Table 7.3. It can be classified according to the VA levels and serves the different application scenarios. Core hardware elements include live data acquisition of process, energy and material data, servers, visualization hardware as well as general lab equipment. The hardware is complemented by a range of software applications. Fig. 7.3 presents impressions of LCDEL at OHLF.

Fig. 7.3 Impressions of the LCDEL at Open Hybrid Lab Factory

7.5   Use Case 1—Life Cycle Engineering in Conceptual Design

The first use case targets the life cycle engineering support of the conceptual design stage for lightweight vehicle bodies (Chap. 2). Therefore, the target audience comprises design engineers as well as project engineers. Both groups of interest have been identified within an initial analysis of typical decision situations in engineering of automotive structures as part of the project MultiMaK2 (Kaluza et al. 2016).

The upper part of Fig. 7.4 (A, B1–3, C) illustrates the engineering context. Design engineering proposes a set of concept alternatives based on given requirements (B1) including different geometries and material combinations (Kaluza et al. 2016). Three geometries of a component cross section (full shape, U-shape, reinforced U-shape) are compared that could be manufactured with different materials and manufacturing processes. Technical parameters like wall thickness can be influenced by engineering design. Mechanical performance of conceptual designs is evaluated, and several alternatives are handed over to an LCA expert (B2) that evaluates scenarios for life cycle environmental performance. Decision-makers, e.g. project managers, need to interpret reports from the domain experts with respect to specific assumptions and scenarios (B3). The lower part of Fig. 7.4 represents an improved engineering process by applying principles of Visual Analytics, in this case realized by implementing a system that integrates LCA modelling and MR visualization (BN1 – BN3). Following this approach, potential trade-offs between design parameters, associated environmental impacts and background scenarios can be determined within ad hoc feedback loops (Kaluza et al. 2019).

Fig. 7.4 Conventional workflow in automotive LCE and concurrent approach based on MR and VA, reproduced from (Kaluza et al. 2019)

Fig. 7.5 illustrates the described engineering situations applying the VA framework. The concept of LCDEL's workflow to support conceptual design will be examined in more detail in the following.

Knowledge layer There are two main goals of enhanced LCDE support in engineering design of automotive body parts through VA.

- Decision-making: Concept alternatives with low environmental impacts should be identified at early stages of conceptual design. Thereby, variability of different fore- and background systems should be considered. Only a small number of conceptual designs should be identified that will be further detailed towards series development.
- Exploration: This task's goal is to enable knowledge gains between the disciplines' engineering processes and thus increase acceptance and effectiveness of suggested LCDE workflows. Therefore, the design space of life cycle environmental impacts and conceptual designs is jointly explored incorporating different materials. For example, what-if scenarios can be performed that show the effect of a parameter variation, e.g. manufacturing yield or process efficiency, to overall life cycle environmental impacts. Other examples would be the analysis of different LCA modelling paradigms or the comparison of different secondary data sources. Exploration incorporates an active engagement of engineers and/ or decision-makers.

Data layer The data layer combines different fore- and background data of a component's life cycle. Primary foreground data is acquired from OHLF manufacturing processes, as described within the following use case (Sect. 7.6). Secondary inventory data is integrated from professional LCA databases, i.e. Ecoinvent or thinkstep GaBi. Another source of secondary data is published information from the state of research. As a large number of innovative manufacturing processes are compared, data availability with respect to expected energy and material demands is typically low. This is especially true

Table 7.4 Visualization portfolio for application in conceptual design

Application	Type	Realization	Knowledge Insights
Joint Concept Engineering	Presentation visualization	Multi-input video wall Display of intermediate engineering results	Elaboration of the design space and boundary conditions Early stage evaluation of conceputla designs
Joint Concept Engineering	Interactive visualization	Mixed reality device (Microsoft HoloLens) 3D models of conceptual designs Structured LCA results	Elaboration of the design space and boundary conditions Early stage evaluation
What-If Analysis	Interactive visualization	Interactive display Web-based or Business Intelligence (Microsoft PowerBI) Integration of external data, e.g. energy, traffic Interface to parameterised LCI model	Contextualization of foreground and background data Identification of trade-offs Narrowing down of design parameters
Concept Selection	Presentation visualization	Single-input video wall Focused selection of few concepts with respect to key parameters Dedicated data visualization libraries, e.g. Seaborn, Vega (Python) + Manual editing	Narrowed down concept comparison Informed decision-making

The Nerve Centre enables a transparent and holistic view on OHLF production. In addition, valuable information can be obtained for further analyses, such as the environmental assessment of hybrid parts. The concept of the Nerve Centre's monitoring system is outlined in the following. As shown in Fig. 7.10, the toolchain of the monitoring system is described by means of the visual analytics framework.

Knowledge Layer As outlined above, the monitoring system intends to support the user in the assimilation of knowledge on the process chain for the manufacturing of hybrid

Fig. 7.6 Joint Concept Engineering—Interactive visualization—Microsoft HoloLens

Fig. 7.7 What-If Analysis—Interactive display

parts through (interactive) visualization methods. The concept of the monitoring system pursues several objectives:

- Decision-making: identification of hotspots at process and factory level (e.g. technical building services) or anomaly detection in process or energy parameters in contrast to a standard behaviour.
- Deeper process analysis: process and data understanding for supplementary deeper cause–effect analysis through the application of machine learning algorithms or the

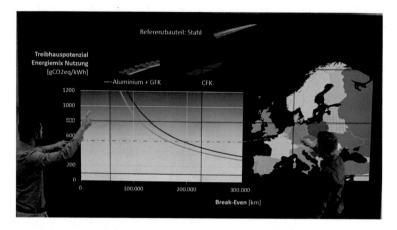

Fig. 7.8 Concept selection—Presentation visualization

Fig. 7.9 Monitoring system at the Nerve Centre of the Open Hybrid LabFactory

usage of the acquired data for parameterization of simulation models (e.g. process and factory simulation).

- Staff development: reduction of entry barrier for students, employees and externals towards the production technologies and intelligent production data analysis.

Data Layer

In order to meet these objectives, the process and energy data of the technical centre is acquired, modelled and transferred into tailored, interactive visualizations in accordance to the visual analytics process. The data layer of the Nerve Centre is composed

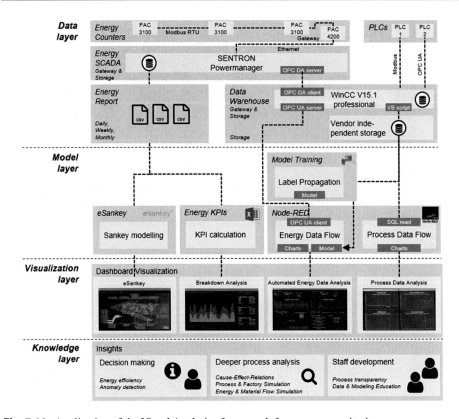

Fig. 7.10 Application of the Visual Analytics framework for process monitoring

of two different data sources. Firstly, energy metering is done through 73 SENTRON PAC energy meters (PAC 3100, 3200 or 4200). The PAC 4200 m serves as an Ethernet-capable gateway to the SCADA system for energy data, which is implemented in terms of the SENTRON powermanager.[1] The software offers historical data access and live monitoring capabilities. Within the Nerve Centre, the system is mainly used as an OPC DA capable gateway, i.e. OPC DA server, for feeding the data warehouse of the Nerve Centre. The second data source of the data warehouse is machine controllers (PLCs). Dependent on the process, e.g. forming, the PLCs provide specific machine and process data as well as sensor data in a high temporal resolution (milliseconds). Examples of process data are the stamp position and its acceleration. In general, controllers employ vendor-specific communication protocols. For efficient data access of all machines, a gateway was selected that supports a large variety of protocols. In the context of the

---

[1] https://support.industry.siemens.com/cs/document/64850998/powermanager-v3-4-sp1?dti=0&lc=en-WW.

Nerve Centre, WinCC professional was chosen as gateway. In addition to the gateway function, the software also supports a data storage function. This enables access to all acquired data through WinCC client applications. The data that is made accessible by the WinCC gateway (forwarding only in case of value changes) is routed to a MySQL database using a Visual Basic script for persistent storage. The stored data can now be used for further modelling steps.

Model Layer

In order to derive knowledge, the raw data collected is processed within the scope of the model layer. This can be done using various approaches, such as agglomeration of data (e.g. statistics), simulations and their parameterization with real data, as well as using machine learning methods. In the sense of the visual analytics process, however, it is also possible to convert the raw data directly into visualizations, such as time series, without much preparation. Within the framework of the Nerve Centre, for example, this is possible for the collected process data of the machines of the technical centre (see Fig. 7.10—process data analysis). Here, modelling using the open-source software Node-RED only involves reading the data from the database and converting it into graphs. A monthly data export from the SENTRON powermanager is carried out for the analysis of historical energy data of the technical centre. The machine-specific energy data is transferred to an Energy Sankey (e!Sankey of ifu Hamburg) and used to calculate KPIs (energy consumption and energy costs per month and consumer group, e.g. technical centre and technical building services) and plot an energy breakdown. A machine learning use case is implemented by means of a machine state recognition based on energy data (details on this can be found in Chap. 4). The first step is to export a training data set from the MySQL database. Within the course of data pre-processing, this data set is partially labelled with its corresponding machine states. Since a semi-supervised learning algorithm (label propagation) is applied, a labelling of the entire data set is not required. This significantly reduces the manual effort involved in data pre-processing. The model trained by label propagation is then stored and can be read into the Node-RED software by the Node-RED contribution machine learning and deployed with live data. The current machine status of a system is shown as text, time series and status distribution. The live energy data is provided via WinCC through an OPC UA server. Node-RED functions as an OPC UA client. In the machine-specific dashboard, the live energy data is visualized similar to the process data besides the machine status information.

Visualization Layer

Table 7.5 summarizes the dashboard applications with regards to their spatial and temporal scale as well as possible knowledge insights (Figs. 7.11, 7.12, 7.13 and 7.14)

Table 7.5 Dashboard portfolio applied at a large-scale video wall

Application	Spatial scale	Temporal scale	Knowledge insights
Energy Sankey	Factory	Minutes - months	- Consumer-specific energy consumptions - Clustering to consumer groups - Comparison with last periods(s) - Hot spot detection
Breakdown Analysis	Factory	Minutes - monthly	- Grouped time series of energy consumptions - Hot spot and anomaly detection - Aggregated energy & cost KPIs
Energy Data Analysis	Process	Seconds - hours	- Live and time series of power demand - Live, time series & distribution of machine state - Automated live energy & cost KPIs based on machine states
Process Data Analysis	Process	Seconds - hours	- Live and time series of process data

7.7   Summary and Outlook

The chapter presents an approach to enhance Life Cycle Engineering workflows for automotive lightweight body parts based on principles of Visual Analytics. Two engineering scenarios have been explored—the support of the conceptual design stage as well as the development of innovative manufacturing processes. The Life Cycle Design & Engineering Lab (LCDEL) at OHLF objectifies the presented workflows. The LCDEL represents a permanent and evolving research infrastructure with the goal to further incubate and mature engineering methods and tools.

Beyond the presented case studies, LCDEL will serve as a platform for future stages of research on innovative automotive structures in the light of sustainable development. On a component level, this includes methods and tools to support the engineering of structural parts that integrate further functions such as electric, acoustic or

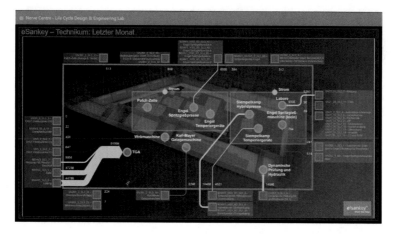

Fig. 7.11 Energy Sankey

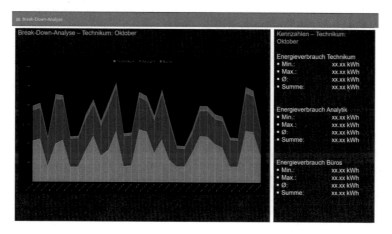

Fig. 7.12 Breakdown Analysis

thermal insulation capabilities. Further, the engineering scope will be broadened towards more systemic perspectives on innovative vehicles and their life cycles. This includes advanced approaches to link component-centred engineering at OHLF with further vivid and highly innovative domains. One example is the joint engineering of structural parts and vehicle drivetrains. Another major focus will lie on the derivation and translation of requirements from adapted vehicle use scenarios and operating models, e.g. changing lifetime distances in mobility-as-a-service.

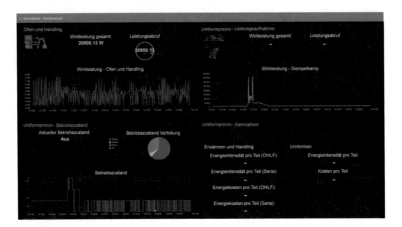

Fig. 7.13 Energy Data Analysis

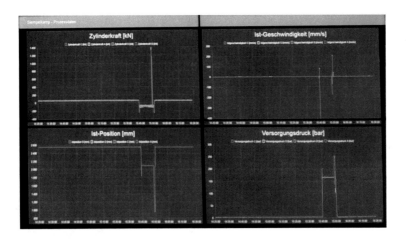

Fig. 7.14 Process Data Analysis

References

Cerdas, F., Thiede, S., Juraschek, M., Turetskyy, A., & Herrmann, C. (2017). Shop-floor Life Cycle Assessment. *Procedia CIRP, 61,* 393–398. doi:https://doi.org/10.1016/j.procir.2016.11.178

Cerdas, F., Thiede, S., & Herrmann, C. (2018). Integrated Computational Life Cycle Engineering — Application to the case of electric vehicles. *CIRP Annals—Manufacturing Technology, 67*(1), 25–28. doi:https://doi.org/10.1016/j.cirp.2018.04.052

Cook, K. A., Earnshaw, R., & Stasko, J. (2007). Discovering the unexpected. *Concrete Construction—World of Concrete, 56*(3), 15–19. doi:https://doi.org/10.1109/MCG.2007.126

Egede, P. (2017). Environmental Assessment of Lightweight Electric Vehicles. Cham: Springer International Publishing. doi:https://doi.org/10.1007/978-3-319-40277-2

Haldin-Herrgard, T. (2000). Difficulties in diffusion of tacit knowledge in organizations. Journal of Intellectual Capital, 1(4), 357–365. doi:https://doi.org/10.1108/14691930010359252

Hauschild, M. Z., Herrmann, C., & Kara, S. (2017). An Integrated Framework for Life Cycle Engineering. Procedia CIRP, 61, 2–9. doi:https://doi.org/10.1016/j.procir.2016.11.257

Kaluza, A., Kleemann, S., Broch, F., Herrmann, C., & Vietor, T. (2016). Analyzing decision-making in automotive design towards life cycle engineering for hybrid lightweight components. Procedia CIRP, 50, 825–830. doi:https://doi.org/10.1016/j.procir.2016.05.029

Kaluza, A., Gellrich, S., Cerdas, F., Thiede, S., & Herrmann, C. (2018). Life Cycle Engineering Based on Visual Analytics. Procedia CIRP, 69, 37–42. doi:https://doi.org/10.1016/j.procir.2017.11.128

Kaluza, A., Juraschek, M., Büth, L., Cerdas, F., & Herrmann, C. (2019). Implementing mixed reality in automotive life cycle engineering: A visual analytics based approach. Procedia CIRP, 80, 717–722. doi:https://doi.org/10.1016/j.procir.2019.01.078

Keim, D., Andrienko, G., Fekete, J., Carsten, G., Melan, G., Keim, D., … Carsten, G. (2008). Visual Analytics : Definition , Process, and Challenges. Information Visualization—Human-Centered Issues and Perspectives, 154–175. doi:https://doi.org/10.1007/978-3-540-70956-5_7

Keim, D. A., Mansmann, F., Stoffel, A., & Ziegler, H. (2009). Visual Analytics. In Encyclopedia of Database Systems (pp. 27–37). Springer. doi: https://doi.org/10.1007/978-0-387-39940-9_1122

Kohlhammer, J., Keim, D., Pohl, M., Santucci, G., & Andrienko, G. (2011). Solving problems with visual analytics. Procedia Computer Science, 7, 117–120. doi:https://doi.org/10.1016/j.procs.2011.12.035

Laurin, L., Amor, B., Bachmann, T. M., Bare, J., Koffler, C., Genest, S., … Vigon, B. (2016). Life cycle assessment capacity roadmap (section 1): decision-making support using LCA. International Journal of Life Cycle Assessment, 21(4), 443–447. doi:https://doi.org/10.1007/s11367-016-1031-y

Li, W., Stanula, P., Egede, P., Kara, S., & Herrmann, C. (2016). Determining the Main Factors Influencing the Energy Consumption of Electric Vehicles in the Usage Phase. Procedia CIRP, 48, 352–357. doi:https://doi.org/10.1016/j.procir.2016.03.014

Rossi, M., Germani, M., & Zamagni, A. (2016). Review of ecodesign methods and tools. Barriers and strategies for an effective implementation in industrial companies. Journal of Cleaner Production, 129, 361–373. doi:https://doi.org/10.1016/j.jclepro.2016.04.051

Thomas, J. J., & Cook, K. A. (2005). Illuminating the path: The research and development agenda for visual analytics. IEEE Computer Society, 184. doi:https://doi.org/10.3389/fmicb.2011.00006

Volkswagen AG. (n.d.). Volkswagen—Mission Statement Environment. Retrieved from https://www.volkswagenag.com/en/sustainability/environment/mission-statement.html (last visited 17.12.2019)

Publications in Course of the MultiMaK2 Project

2019

- Kaluza, Alexander; Juraschek, Max; Bueth, Lennart; Cerdas, Felipe; Herrmann, Christoph. Implementing mixed reality in automotive life cycle engineering: A visual analytics based approach. In: Procedia CIRP 80, Elsevier B.V., Amsterdam, 2019, 717–722, 10.1016/j.procir.2019.01.078

2018

- Kaluza, Alexander; Gellrich, Sebastian; Cerdas, Felipe; Thiede, Sebastian; Herrmann, Christoph. Life cycle engineering based on visual analytics. In: 25th CIRP Life Cycle Engineering (LCE) Conference, Copenhagen, Denmark, Procedia CIRP 69, Elsevier B.V., Amsterdam, 2018, 37–42, 10.1016/j.procir.2017.11.128
- Herrmann, Christoph; Dewulf, Wim; Hauschild, Michael; Kaluza, Alexander; Kara, Sami; Skerlos, Steven. Life cycle engineering of lightweight structures. In: CIRP Annals - Manufacturing Technology, Elsevier B.V., Amsterdam, 2018, 67, Issue 2, 651–672, 10.1016/j.cirp.2018.05.008

2017

- Kaluza, Alexander; Kleemann, Sebastian; Fröhlich, Tim; Vietor, Thomas; Herrmann, Christoph. Concurrent design & life cycle engineering in automotive lightweight component development. In: Procedia CIRP 66, Elsevier B.V., Amsterdam, 2017, 16–21, 10.1016/j.procir.2017.03.293
- Kleemann, Sebastian; Fröhlich, Tim; Türck, Eiko; Vietor, Thomas (2017): A Methodological Approach Towards Multi-material Design of Automotive Components. In: Procedia CIRP 60, 68–73. 10.1016/j.procir.2017.01.010

© Springer-Verlag GmbH Germany, part of Springer Nature 2023
T. Vietor (ed.), *Life Cycle Design & Engineering of Lightweight Multi-Material Automotive Body Parts,* Zukunftstechnologien für den multifunktionalen Leichtbau, https://doi.org/10.1007/978-3-662-65273-2

- Fröhlich, Tim; Kleemann, Sebastian; Türck, Eiko; Vietor, Thomas (2017): Multi-criteria analysis of multi-material lightweight components on a conceptual level of detail. 10.24355/DBBS.084-201709070942
- Kleemann, S., Inkermann, D., Bader, B., Türck, E., & Vietor, T. (2017). A Semi-Formal Approach to Structure and Access Knowledge for Multi-Material-Design. In 21st International Conference on Engineering Design (ICED17), 21–25 August 2017. 10.24355/dbbs.084-201708301114

2016

- Kaluza, Alexander; Kleemann, Sebastian; Broch, Florian; Herrmann, Christoph; Vietor, Thomas. Analyzing Decision-making in Automotive Design towards Life Cycle Engineering for Hybrid Lightweight Components. In: 26th CIRP Design Conference, Stockholm, Elsevier B.V., 2016, 825–830, 10.1016/j.procir.2016.05.029
- Kleemann, S., Türck, E. & Vietor, T. (2016). Towards knowledge based engineering for multi-material-design. In: DS 84: Proceedings of the DESIGN 2016 14th International Design Conference, 2027–2036

2015

- Cudok, Anja; Hasenpusch, Jan; Inkermann, David; Vietor, Thomas: Vorstellung einer Methodik zur Identifikation von Bauteilen mit Potential zur Gestaltung in Hybridbauweise. In: DFX 2015: Proceedings of the 26th Symposium Design for X, 255–266

Printed in the United States
by Baker & Taylor Publisher Services